Particle Acceleration and Detection

The series "Particle Acceleration and Detection" is devoted to monograph texts dealing with all aspects of particle acceleration and detection research and advanced teaching. The scope also includes topics such as beam physics and instrumentation as well as applications. Presentations should strongly emphasize the underlying physical and engineering sciences. Of particular interest are

- contributions which relate fundamental research to new applications beyond the immediate realm of the original field of research
- contributions which connect fundamental research in the aforementioned fields to fundamental research in related physical or engineering sciences
- concise accounts of newly emerging important topics that are embedded in a broader framework in order to provide quick but readable access of very new material to a larger audience

The books forming this collection will be of importance to graduate students and active researchers alike

More information about this series at http://www.springer.com/series/5267

Thomas Otto

Safety for Particle Accelerators

 Springer

Thomas Otto
CERN
Genève, Switzerland

ISSN 1611-1052   ISSN 2365-0877   (electronic)
Particle Acceleration and Detection
ISBN 978-3-030-57030-9   ISBN 978-3-030-57031-6   (eBook)
https://doi.org/10.1007/978-3-030-57031-6

This Springer imprint is published by the registered company Springer Nature Switzerland AG
The registered company address is: Gewerbestrasse 11, 6330 Cham, Switzerland

Preface

Particle accelerators have received ever growing attention since their invention in the 1930s. They have developed from one-of-a-kind facilities for fundamental research to a wide range of facilities with specialized applications. Their use in fundamental research is no longer limited to nuclear and particle physics. Synchrotron light sources are, despite their diminutive name, full-fledged particle accelerators and serve material science, chemistry, and biology. Spallation neutron sources employ the most powerful proton accelerators (measured by the product of beam intensity and maximal beam energy) so far constructed. Medicine employs scores of compact electron linear accelerators with a limited energy range (E < 25 MeV), and the advent of proton and heavy ion therapy has led to the construction of specialized synchrotron facilities. Industry uses accelerators, for example, in the production of semiconductors, for the sterilization of medical equipment, and for the polymerization of plastics.

The extension of fundamental research facilities to large national or transnational organizations led to a simplified access of researchers to accelerators. These users are not specialized in accelerator technology and related fields, but they have diverse academic and technological backgrounds. The entry of accelerators into the medical field enhances the demand for reliable and safe operation in the interest of patients and medical staff alike. In industry, finally, the emphasis is on reliability and cost-effectiveness. In short, accelerators are used by many non-specialists as a tool which must conform to regulations and standards like other means of production.

International bodies and national regulators define stringent technical and organizational criteria for the safe use of technical facilities and equipment, targeted to protect clients, users, and the environment alike from harm. These regulations apply naturally to particle accelerators. Operators and users are legally obliged to implement regulations and morally bound to respect recommendations as much as technically feasible.

Occupational safety is concerned with controlling the hazardous effects of professional activities for workers, the public, and the environment. Occupational safety draws essential information from physics, chemistry, engineering, occupational health, and environmental protection for providing an optimal, appropriate protection from occupational hazards. For particle accelerators, occupational safety creates the conditions for obtaining research, medical, or business objectives in a safe and responsible manner without creating harm for workers, the public, and the environment.

This book gives an overview of the vast subject of safety at particle accelerators. It is targeted at managers, scientists, engineers, and users at accelerator facilities. It serves also as a first introduction for safety professionals who take up work at a particle accelerator or who are generally curious on safety in these facilities. The book is organized into five chapters. Chapter 1 gives a working definition of hazard and risk as an essential introduction to occupational safety. Chapter 2 briefly introduces particle accelerators and then describes safety aspects of their core technologies: magnets, cryogenics, radiofrequency, lasers, and beam interception. Chapter 3 treats safety topics connected to accelerator beams and ionizing radiation. In Chap. 4, safety hazards also occurring in major industries are described in their relation to particle accelerators: electrical and mechanical safety, pressure vessels, fire safety, occupational noise, and environmental impact. The concluding Chap. 5 closes the loop opened in Chap. 1 with a more detailed description of the safety process, it gives an overview of safety organization in accelerator centres, and describes beam safety and functional safety, a concept frequently employed at accelerators.

At around 150 pages, the book cannot be exhaustive, but presents an overview of the subject to the interested reader. References are given to recent literature, preferably to documents freely available on the Internet or as links to websites. All references were up to date at the time of publication; however, occupational safety is a rapidly evolving field. References to regulatory context (directives, laws, standards) may change rapidly, and the reader is warned to check all references for validity before applying them in a real-world accelerator facility.

Safety is based on science and engineering, but it is not a hard science. The "human factor" and the personal viewpoint inevitably enter in risk assessments and in the implementation of mitigation measures. I regularly profit from the collaboration with fellow occupational safety specialists when defining the best risk control measures within a limited budget of money, time, and manpower. The responsibility for this book's content, including any mistakes, is nevertheless solely my own. The opinions expressed in the book do not always reflect policies of my employer, the European Organization for Nuclear Research, CERN.

This book is dedicated to Marta, my wife and greatest support. Thank you!

Stay safe and healthy.

Geneva, 2020   Thomas Otto

Acknowledgements

A picture says more than a thousand words, and I thank all organisations and companies who allowed me to reproduce their photographs and illustrations in this book:

- CERN and PSI for the right to use several photographs of their facilities.
- IBA in Louvain-La-Neuve (BE) (Fig. 2.1).
- The European Commission's Directorate-General for Employment and Social Affairs (Fig. 2.21).
- The International Electrotechnical Commission (IEC) for permission to reproduce information from its International Standards (Fig. 2.23). IEC has no responsibility for the placement and context in which the extracts and contents are reproduced by the author, nor is IEC in any way responsible for the other content or accuracy therein.
- Lawrence Berkeley National Laboratory (USA), D.E. Groom, S.R. Klein, for Fig. 3.1 from the Review of Particle Physics.
- RadPro International GmbH in Wermelskirchen (DE) (Figs. 3.7 and 3.8).
- Public Health England (PHE) (Fig. 3.9).
- ELSE NUCLEAR S.r.l. in Busto Arsizio (IT) (Fig. 3.10).
- The Swiss Occupational Accident Insurance SUVA in Luzern (CH) (Fig. 4.2).
- The UK Health and Safety Executive (HSE) (Fig. 4.15). This Figure is published under Open Government License (OGL), https://www.nationalarchives.gov.uk/doc/open-government-licence/version/3/.

Finally, a big thank you goes to Hisako Niko from Springer, who accompanied this book from the beginning and who was always very patient with my requests for yet another delay.

Contents

Chapter 1
Introduction to Occupational Safety

Abstract It is not possible to write a book on Safety at Accelerators, or on occupational safety in general, without a practical understanding of the technical language of the field. Like in other scientific or technical fields, words are used with a specific meaning, often deviating from their everyday use. These terms must be clarified before penetrating deeper into the subject. Central to occupational safety are the terms of hazard, risk, and control. A more detailed description of these concepts is found in Chap. 5.

1.1 Hazard

An informal definition of *hazard* is simply "something with the potential to cause harm". In occupational safety, harm manifests itself either by an accident or an occupational illness.

Hazard can be classified by the technical domain it originates from (e.g. mechanical, electrical, chemical, physical, psychosocial hazards). Another way of characterisation is by the vector of the hazard: the potential harm is carried by an *equipment* (for example machine, tool, experimental apparatus), an *activity* (for example production, construction, maintenance, dismantling) or a *substance*, either used in the activity (basic material, catalyser, …) or produced in the process (exhaust gases, dust, …). Numerous hazards exist at the workplaces in an accelerator facility: in the accelerator building or tunnel, in workshops, laboratories and in offices. This book puts the focus on specific hazards at particle accelerators, for example magnetic fields, cryogenic temperatures, and ionising radiation. Industrial hazards present in industry and services impact also the construction, operation, and maintenance of accelerators and one chapter is dedicated to them with references to guidelines in occupational health and safety (OHS).

A useful tool to classify hazards is a hazard list, as reproduced in Annex A. It contains both general hazards from industry and business and specific hazards occurring at accelerator sites.

© The Author(s) 2021

T. Otto, *Safety for Particle Accelerators*, Particle Acceleration and Detection,
https://doi.org/10.1007/978-3-030-57031-6_1

1.2 Risk

The informal definition of *risk* is "a measure of the probability of a hazard to cause harm, and of the severity of the consequences." By reducing risk, one reduces the probability or/and the severity of harmful effects and therefore improves safety.

Measure is often meant in a qualitative sense. Risk can be classified on a coarse scale as low, medium, or high. Such a qualitative judgement is based on a good knowledge of the employed activities, equipment and substances and previous professional experience and remains always subjective.

Quantitative risk assessment makes use of published data on failure rates of components or equipment, accident rates and of models for the behaviour of complex systems. It is a specialist's domain and is employed in high-hazard industries, such as the chemical and nuclear industries [2].

Often a risk assessment will be situated between the purely qualitative and quantitative and it is up to the professional judgement of the safety specialist and the line management to determine the correct level of assessment.

Risk and hazard are often confounded in colloquial language. People speak of a "high risk" activity, but they usually mean a high hazard activity. While there may exist considerable hazards at a workplace, it is the purpose of occupational safety to reduce and control the risk of these hazards in such a way that workers are not harmed and that the integrity of the public and the environment is preserved.

1.3 Control

The goal of occupational safety is the elimination of hazards and the diminution of risks. One speaks in this context of "control" (or control measure). To control a danger at the workplace, one may progress by elimination, install technical protection measures, equip the worker with personal protections or instruct and train them to avoid the hazards. In many cases, the required control for a given hazard or risk is dictated by law and regulations. Where this is not the case, it is up to line management and occupational safety specialists to agree on the appropriate level of control. It shall reduce risk to a level acceptable by management and workers, without impeding the activity (e.g. the operation of a facility, the conduction of an experiment) beyond the necessary, and without causing incommensurate cost. Choosing and implementing an appropriate level of control is a matter of professional judgement, and as such is based on experience.

The European Union has published a directive, a document that the member states must translate in national law, where a hierarchy of controls is first evoked [1]. The directive stipulates that the employer shall take the measures necessary for the safety and health protection of workers. He or she shall implement these measures based on the following general principles of prevention:

(a) avoiding risks.
(b) evaluating the risks which cannot be avoided
(c) combating the risks at the source
(d) adapting the work to the individual, especially as regards the design of work places, the choice of work equipment and the choice of working and production methods, with a view, in particular, to alleviating monotonous work and work at a predetermined work-rate and to reducing their effect on health
(e) adapting to technical progress;
(f) replacing the dangerous by the non-dangerous or the less dangerous;
(g) developing a coherent overall prevention policy which covers technology, organization of work, working conditions, social relationships and the influence of factors related to the working environment;
(h) giving collective protective measures priority over individual protective measures;
(i) giving appropriate instructions to the workers.

This European directive defines a framework for occupational health and safety, based on a few principles. More than 30 years old, it has shown its effectiveness and has not been revised since. Its principles influence all following directives and laws relating to occupational safety. The U.S. National Institute of occupational safety and health has cast the hierarchy of controls in a more useful way [3]. It gives preference to eliminating or replacing a risk, and to collective protective measures. Personal protective equipment, technical instruction and safety training of the workers is designated as the least effective measure in this hierarchy of controls. It is known that the effectiveness of these measures for the reduction of accidents and occupational illness depends on individual factors and is generally lower than the technical measures listed first.

References

1. Council Directive of 12 June 1989 on the introduction of measures to encourage improvements in the safety and health of workers at work, 89/331/EEC, http://data.europa.eu/eli/dir/1989/391/2008-12-11
2. Directive 2012/18/EU on the control of major-accident hazards involving dangerous substances, http://data.europa.eu/eli/dir/2012/18/oj
3. National Institute for Occupational Safety and Health (NIOSH), Hierarchy of Controls, https://www.cdc.gov/niosh/topics/hierarchy/#

Chapter 2
Risks and Hazards of Particle Accelerator Technologies

Abstract In this section, the motivation and operation of particle accelerators are briefly introduced. Then, safety aspects of the key building blocks are treated. Magnets provide the steering forces for accelerated particles. Cryogenics provides the low temperatures required for the operation of superconducting magnets; radio-frequency technologies impart energy to accelerated particles. A byproduct of their operation is Non-ionising radiation. Another type of NIR is represented by lasers which find increasing use in accelerator applications. Finally, collimators shape the particle beams and protect sensitive elements, while dumps absorb the particles at the end of their course.

2.1 Accelerators for Pedestrians

2.1.1 Why Particle Accelerators?

Fundamental research provided the first motivation to accelerate subatomic particles to ever higher energies. One approach to understand the need for high energies is the analogy to an optical microscope: to analyse the properties of subatomic particles, with dimensions less than a femtometre (1 fm = 10^{-15} m), probes with a wavelength of the same order of magnitude were required. The de Broglie wavelength λ_B of a relativistic particle with momentum p is:

$$\lambda_B = \frac{h}{p} = \frac{hc}{\sqrt{E^2 - \left(mc^2\right)^2}} \cong \frac{hc}{E} \tag{2.1}$$

The approximation is valid in the ultrarelativistic case where the total energy E is at least three times the rest-mass energy mc^2 of the particle. Inserting numerical values in this equation yields an energy of 1.24 GeV, which is ultrarelativistic for electrons, to achieve a De Broglie wavelength of 1 fm. For probing smaller structures, even higher energies would be required.

A second picture to grasp the need for high energies is the creation of new particles. Einstein's relation of the equivalence of matter and energy is:

© The Author(s) 2021   5
T. Otto, *Safety for Particle Accelerators*, Particle Acceleration and Detection,
https://doi.org/10.1007/978-3-030-57031-6_2

$$E = m_0 c^2$$

To generate a particle with rest mass m_0 in a relativistic collision, the total centre-of-mass energy of the colliding particles must be at least of the value indicated by Einstein's relation. In CERN's electron-positron collider LEP, operating in the 1990s, the particles had an individual energy of 45 GeV [2]. In a head-on collision this added to 90 GeV, enough to produce the Z-Boson, one of the mediators of the electro-weak interaction. Much higher particle energy is required when particles collide with a fixed target, for this reason the high-energy frontier of particle physics is explored by colliders.

Ever higher energies than achievable by LEP were necessary to test theoretical models of particle physics, a process which culminated so far with the construction and operation of CERN's LHC at a centre-of-mass energy of 13–14 TeV [3]. It permitted the discovery of the Higgs boson, the last missing element in the Standard Model of particle physics. Its present objective is to find evidence of physical processes not described by the Standard Model, and to lay the experimental fundament to a new, more complete theory of particle physics.

2.1.2 The Particle Accelerator Family

The simplest accelerator was found in cathode ray tubes (in wide use for monitors in television sets and oscilloscopes until the breakthrough of flat screens). Electrons were emitted by a cathode (a heated piece of metal), accelerated by a static voltage gradient between cathode and anode, and deflected by transversal electric or magnetic fields. The energy of an electron with charge e traversing an electrical voltage difference U is $E = eU$, it is expressed in the unit eV (1 eV = 1.602 10^{-19} J). The accelerated electrons hit a phosphorescent screen, thus forming a visible image. The limitation of a static voltage accelerator is the voltage breakdown between anode and cathode.

2.1.2.1 Linear Accelerator

Linear accelerators ("linac") overcome the limitation of the breakdown voltage by accelerating particles passing through a series of aligned accelerating structures. This allows adding up the energy gain. Linear accelerators are relatively simple and are the workhorse in the medical field, where they accelerate electrons to energies between 6 and 25 MeV for the generation of bremsstrahlung X-rays for diagnostics and therapy. An estimate places the number of medical linacs at 14000 worldwide, counting for approximately 30% of all accelerators.

Modern high-power accelerators are realised as linear accelerators. A recent, operating example is the Spallation Neutron Source SNS at Oak Ridge (USA). A liquid mercury (Hg) target is bombarded by a proton beam with an energy $E = 1$ GeV and a beam power $P = 1.4$ MW. Mercury atoms are shattered upon impact of a proton and 20–30 neutrons are released per collision. They are moderated and guided to experimental stations with spectrometric instruments. More power translates into more neutron flux, and the SNS plans an upgrade to $E = 1.3$ GeV and $P = 2.8$ MW [8]. The European Spallation Source ESS in Lund (SE), presently under construction, will use a proton driver with $E = 2$ GeV and $P = 5$ MW [5].

Similar instantaneous beam intensities as for spallation neutron sources are envisaged for beam-dump facilities. In these, the accelerated particle beam is projected onto a massive beam dump/ target. Detectors are placed downstream from the beam dump /target, where it is hoped that exotic, hitherto unobserved particles can be identified.

At the high-energy end, linear accelerators become very long. The International Linear Collider Study (ILC) projects a linear electron/positron collider which could be built in several stages, with a final collision energy of 500 GeV. At this energy, the two Linacs built in opposing direction would have a length of 31 km [7].

2.1.2.2 Cyclotron

In a linac, the accelerated particles pass each accelerating gap only once. In a circular accelerator, of which the simplest example is the cyclotron (Fig. 2.1), the accelerating gap is passed repetitively. Between the poles of a large electromagnet, two D-shaped, hollow electrodes are placed. Between the "Dees", as these electrodes are called, a high-frequency alternating voltage is applied. In the centre of the slit

Fig. 2.1 Compact cyclotron IBA Cyclone® KIUBE for production of radiopharmaceuticals by protons with kinetical energy up to $E_{kin} = 18$ MeV. (Image: IBA, Louvain-la-Neuve, BE)

separating the "Dees" an ion source is placed. Ions emitted by the source will be accelerated by passing the field gap between the "Dees", while their path is bent by the magnetic dipole field. At every passage they gain kinetic energy, leading to the next orbit with a higher radius because magnetic flux density B, particle momentum p and orbit radius ρ are connected by the relation

$$B\rho = \frac{p}{q} = \frac{pc}{qc} \approx \frac{E}{qc}.$$

Once they have passed the gap between the Dees, the particles are shielded by the Dee-walls from the electrical field which passes through the reverse polarity. At the outer radius of the "Dees", the particles are extracted at their maximum energy by a septum magnet.

There are two limitations to the cyclotron principle:

• At high energies, relativistic effects become important and the revolution frequency in the magnetic field is no longer matching the frequency of the electrical accelerating field. This can be overcome by introducing variable accelerating frequencies in so-called synchrocyclotrons.
• The maximal radius of the particles' orbit and thus the maximal energy is determined by the size of the magnetic poles. The presently largest cyclotron is located at TRIUMF in Vancouver (CA), its magnet has a diameter of 18 m and a magnetic flux density of 0.46 T. The mechanical problems of a large magnet size can be overcome by splitting it in several units. The Swiss Paul Scherrer Institute in Villigen (CH) operates its ring cyclotron with eight sectors and obtains a beam power of $P = 1.2$ MW at an energy of $E = 590$ MeV [6].

Smaller Cyclotrons are used in medical centres to produce radioisotopes from which radiopharmaceuticals are synthesised for diagnostics (SPECT and PET) and for targeted tumour therapy. In industry, cyclotrons are used to implant ions into materials to modify their physical properties, for example in highly integrated microelectronic circuits.

2.1.2.3 Synchrotron

To reach even higher particle energies, one resorts to repetitive acceleration in a synchrotron, another type of circular accelerator. Here, the average particle orbit is closed, the path of the particle oscillates around it. The magnetic bending field is produced by dipole magnets arranged along the particle orbit; their field extends only over the comparatively small volume of the beam line. The magnetic flux density is increased synchronously with the particle's energy gain so that the average radius of the orbit remains constant.

Accelerators with the synchrotron principle are built from circumferences of a few 10 metres to 27 km (CERN's LHC, [3]), and synchrotrons with a circumference of nearly 100 km are envisaged (CERN's Future Circular Collider, [4]).

Synchrotrons are used predominantly in research applications, where particle beams are accelerated and collided either with fixed targets or with each other (see below, "Collider"), or produce synchrotron radiation (see below, "Storage Ring"). A few medical treatment facilities world-wide employ synchrotrons for accelerating protons or charged heavy ions for hadron therapy. For a few cancer types and sites, this modality of radio oncology is advantageous over conventional radiotherapy and requires accelerators of a size comparable to those of research centres.

2.1.2.4 Storage Ring

The first synchrotrons were built to accelerate particles to high energies and to make them collide against external or internal targets as soon as they reached the terminal energy.

There are two reasons to keep particles for extended times circulating in a synchrotron, which is then called a *storage ring*:

- The generation of synchrotron radiation
- The collision of particles with each other (see below, "Collider")

Synchrotron radiation is a by-product of the acceleration and change of direction of electrons. The electromagnetic radiation emitted upon momentum change is in the near-UV range of wavelengths and is used for material and biological research. While one tries to minimise the emission of synchrotron radiation in high-energy accelerators, because of the induced loss of energy, one maximises it in special circular accelerators, the synchrotron light sources (Fig. 2.2). They prove invaluable in material research, from semiconductors to biological samples.

Two features of synchrotron light sources are:

Fig. 2.2 Swiss Light Source SLS at PSI, Villigen, a synchrotron light source. The round accelerator building has an external diameter of 138 metres. (Images reproduced with permission by: Paul Scherrer Institut)

- the emission of synchrotron radiation is enhanced by making the particles follow undulatory paths in straight sections of the ring (so-called Undulators and Wigglers).
- the energy emitted in form of synchrotron light must be continuously restored by operating the accelerating RF cavities.

In synchrotron light sources, the electrons may be stored in the synchrotron which accelerated them in the first place. This leads to an operation scheme where the ring is filled with a low-energy beam provided by a linac or a smaller synchrotron, the electrons are accelerated to the terminal energy, and then synchrotron light is produced and emitted. The light intensity will decrease over time because of beam loss. A more flexible arrangement is to separate the synchrotron for acceleration and the storage ring. As soon as the intensity of the emitted synchrotron light become too weak one can "top up" the filling of the storage ring with a small injection form the accelerator synchrotron. This mode is called "top-up operation". The split between accelerator and storage ring has other advantages. For example, the magnets in the storage ring need to operate only in a very small range around the maximal energy and thus can have a better field quality.

2.1.2.5   Free-Electron Laser

Free-Electron Lasers (FEL) are a specific type of synchrotron light source, in which the undulating structure of the beam path emits coherent photons (i.e. Laser, Sect. 2.5) in the low-energy X-ray range. This is the only known source of X-ray lasers and opens a new research field exploiting the spatial and temporal coherence of the radiation. FEL are built either as a linear accelerator or in "racetrack" configuration, which one can imagine as a coiled-up linac in which the same accelerating cavity can be used during several passages.

2.1.2.6   Collider

A collider is the combination of two storage rings[1] in which the beams circulate in opposite directions and are brought to collide at defined locations, in the centre of large particle physics detectors. A collider combines the characteristics of an accelerating synchrotron and a storage ring. It consists of circular arcs which are connected by long straight sections in which the collision points are inserted. Special focussing magnets provide for the charged particle optics to bring the beams to

---

[1] In special cases, a collider can be realized within a single storage ring when the colliding particles are antiparticles to each other: Examples are the SPS (proton-antiproton) and the LEP (electron-positron), both at CERN. CERN's LHC looks like a single ring, but in reality, the magnet yoke houses two magnets with opposite polarity.

collision in the smallest possible volume, to enhance the probability of a collision, and to bring them back to a stable orbit after the collision point.

2.1.3 Particle Acceleration from Source to Target

After this overview of different particle accelerator types, I describe the path of a particle from its source to the collision with another particle. This serves to introduce different types of hardware which are subject of the following sections.

Charged particles are generated in a *particle source*, for a proton accelerator, hydrogen atoms coming from an ordinary gas bottle are ionised in a plasma ion source. Electrons are emitted by a cathode, either by heating it, or by shining laser light (Sect. 2.5.1) onto it. The second method allows to impregnate a time structure on the flux of emitted electrons.

In their further parcourse, the charged particles are manipulated by electromagnetic fields, one can distinguish between

- accelerating fields, exploiting the potential difference between two electrodes or in a time-varying radiofrequency field (Sect. 2.4) to exert an accelerating force on the charged particle, and
- guiding fields generated by electromagnets (Sect. 2.2), changing the direction of the particle's path.

In a simplifying manner one can say, that in a linear accelerator the technology defining the accelerator performance is radiofrequency (RF) technology, while in a circular accelerator it is magnet technology.

In a linear accelerator, the particle passes a series of RF cavities in which it draws kinetic energy from the electromagnetic field. The most efficient RF cavities are superconducting, requiring cryogenic technology (Sect. 2.3) to reach their operating temperature. Quadrupole magnets between the cavities keep the beam well focused. One the particle has reached the terminal energy of the accelerator, it is used to produce X-rays (in a medical linac), neutrons (in a neutron spallation source) or coherent X-rays (in a free electron Laser).

In a circular accelerator, the beam passes the same RF cavity repeatedly. It either spirals outward with increasing energy (Cyclotron) or it is kept on a unique orbit by magnets with increasing flux density (Synchrotron). To achieve the highest particle energies, strong magnetic fields are provided by superconducting magnets (Sect. 2.2.2) cooled to operating temperature by cryogenic technology (Sect. 2.3).

Having reached the highest possible energy of a synchrotron, the beam is sent on a target (Sect. 2.6.2), for production and subsequent investigation of other, exotic particles, for example antiprotons or unstable nuclei.

Alternatively, in a storage ring or a collider, the beam is left circulating at the highest energy. Here, dissipative processes make the beam size larger. These processes include scattering on rest-gas atoms in the evacuated beam line and electromagnetic and strong interaction during near-collisions with the opposing beam particles, To prevent excessive beam loss in sensitive equipment, collimators (Sect. 2.6.1) remove particles which have moved too far away from the ideal orbit. Once the beam intensity has become too low to produce a satisfying rate of collisions or intensity of synchrotron light, the beam is deflected by switches into large absorbers, so-called beam dumps (Sect. 2.6.3), where the energy is absorbed. If large deviations of the particle's path from the nominal orbit are observed, an emergency beam dump is triggered automatically to prevent damage to accelerator elements by a massive impact of the particle beam.

2.2 Magnets

Magnetic fields guide and focus particles on their orbit in circular and linear particle accelerators (Fig. 2.3).

One can distinguish between dipole magnets, generating the field necessary for bending the beam, quadrupole magnets for focussing it, and higher-order multipole magnets for correcting either deviations from the ideal field shape by the simpler magnets, or non-linear ion-optical effects induced by the particle beam itself.

2.2.1 Normal Conducting Magnets

A normal conducting magnet is an electromagnet powered by a resistive current (running in "normal" conductors), operating at room temperature. It consists of coils to produce the magnetic induction, which are wound around an iron yoke with the purpose of shaping the magnetic flux lines to the desired multipole shape. As a rule of thumb, normal conducting magnets can be used for flux densities of up to $B = 2$ T, above this value, only superconducting magnets are economically viable.

Dipole magnets create a homogeneous magnetic field in the gap of their yoke, oriented at a right angle to the direction of the particle beam. The Lorentz Force deflects the beam in the direction orthogonal to both its original velocity and the

Fig. 2.3 Warning sign for magnetic field hazards, after [1]. (Source: https://publicdomainvectors.org)

Fig. 2.4 Schematic cut through a C-Type Dipole magnet in direction of the beam. l_{iron} and l_{air} are the length of the magnetic flux lines in the yoke and in the air gap, respectively. (From [9])

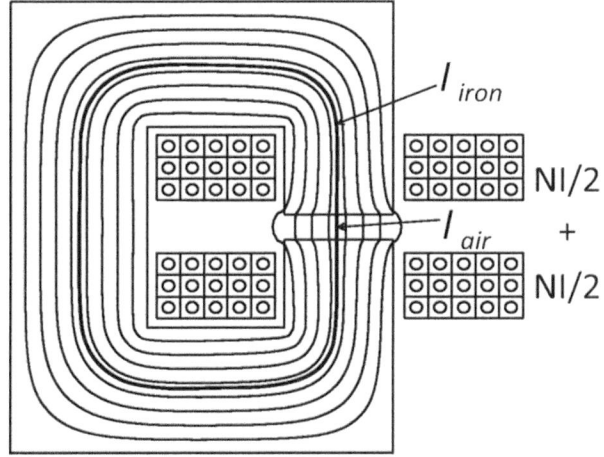

magnetic field lines. A popular realisation of a dipole magnet is the C-type magnet (Fig. 2.4), because it leaves easy access to the vacuum chamber enclosing the particle beam.

The magnetic flux density in the gap of a C-type dipole magnet is given by

$$N \cdot I = \eta \frac{B}{\mu_0} l_{air}$$

Here, N is the number of turns of the electrical conductor, I is the electrical current, l_{air} is the gap-width between the magnet poles and η is an efficiency coefficient and usually close to one. The required magnetic flux density can be achieved by varying the number of turns or the electrical current.

The energy stored in the magnet is

$$W_m = \frac{1}{2} L I^2$$

with the inductivity given by $L = \eta \mu_0 N^2 A / g$. The shape factor A is related to the geometry of the magnet, in a dipole it corresponds approximately to the product of length and width of the magnet pole. Further information can be found in [9, 16].

As an example, the parameters of a dipole magnet in the Super Proton Synchrotron SPS are given in Table 2.1 and its layout is illustrated in Fig. 2.5. This magnet is a Window-frame magnet, for which the relation between current-turns, gap-width and magnetic flux density is, in first order, identical to a C-type magnet.

Table 2.1 Parameters of SPS
dipole magnet type B2

Peak momentum	pc	450 GeV
Bending radius	ρ	741 m
Required magnetic flux density	B	2.04 T
Gap width	g	52 mm
Number of turns	N	16
Peak current	Ipeak	5.75 kA
Coil resistance (1 magnet)	R	4.4 mΩ
Inductance (1 magnet)	L	9.9 mH
Stored energy at peak current (1 magnet)	W	163 kJ

Source: [11], p. 47 and [10]

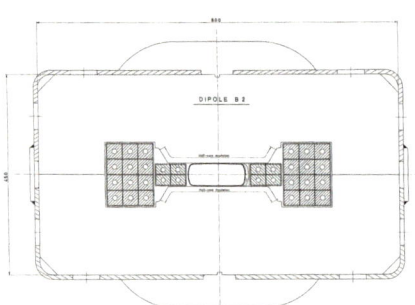 

Fig. 2.5 SPS Dipole magnet Left: Cut through the magnet [11]. Right: magnet on a transport carriage. The closed current loops are clearly visible in this perspective. (Copyright CERN, reused with permission)

2.2.2 Superconducting Magnets

Normal conducting magnets with iron yokes are limited to a magnetic flux density of approximately 2 T because of saturation effects in the iron. Stronger magnetic fields can be produced in superconducting magnets. Superconductivity is a low-temperature phenomenon in many metals, where electrical resistance R vanishes once the material has a temperature $T < T_c$, the critical temperature.

Superconductivity was discovered by H. Kammerlingh Onnes in 1911 in mercury below the critical temperature of $T_c = 4.2$ K, after having liquefied helium three years earlier. It was soon discovered that superconductivity is also a function of magnetic flux density B and current density J, beyond a critical value B_c or J_c the material becomes normal conducting. The boundary between super- and normal conducting state can be described by the relation of the critical current J_c as a function of temperature T and magnetic flux density B and visualized as a hyperplane in the 3-dimensional phase diagram with coordinates T, B and J (Fig. 2.6).

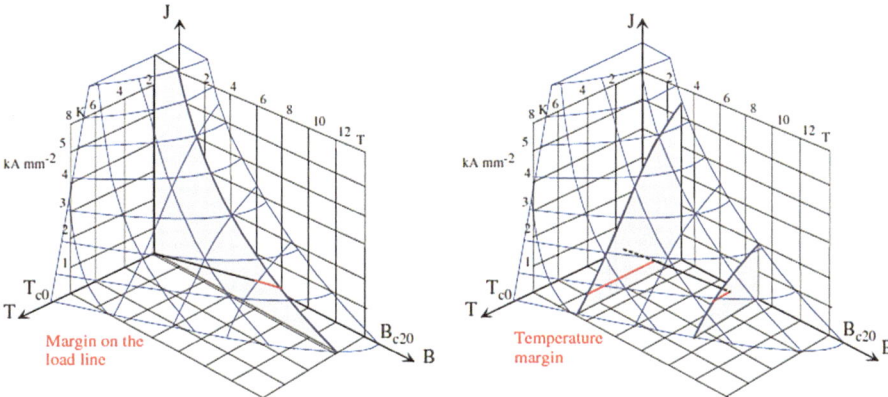

Fig. 2.6 Critical surface of the Nb–Ti superconductor [14]. The critical surface separates points in T-B-J space in which the magnet is superconducting from those where it is normal conducting. Left: Operational margin of the magnet on the load line, i.e. how much J or B may increase before the magnet becomes normal conducting. Right: Temperature margin at a given field and current density, i.e. how much T may increase before the magnet becomes normal conducting. (Image reproduced with permission by John Wiley and Sons)

Table 2.2 Parameters of LHC dipole magnet

Peak momentum	pc	7000 GeV
Bending radius	ρ	2800 m
Required magnetic flux density	B	8.33 T
Aperture (Gap width)	g	53 mm
Nominal current	I_{peak}	11.9 kA
Inductance (1 magnet)	L	98 mH
Stored energy at nominal current (1 magnet)	W	6.93 MJ

Source [17]

The presently most used superconducting alloy for magnets is Nb-Ti, with a critical temperature of 9.2 K at zero field and zero current. To achieve a magnetic flux density of 8 T with sufficient operational margin for accelerator operation it is necessary to cool the magnet to a temperature of approximately 1.9 K.

Table 2.2 and Fig. 2.7 illustrate the LHC dipole magnet, the most powerful accelerator magnet currently in operation. Higher flux densities of up to 16 T can be reached by using the alloy Nb$_3$Sn, with a critical temperature of 18.2 K at zero field and current density.

Fig. 2.7 LHC Dipole magnet. Left: cut through the magnet [17], Right: artist's 3D view of an opened dipole in the tunnel. (Copyright CERN, reused with permission)

2.2.3   Safety Aspects of Magnets

The nature of hazards from normal and superconducting magnets are similar: exposure to strong magnetic fields, exertion of magnetic forces on metallic objects and electrical hazards. The hazards emerging from the use of cryogenics will be treated in Sect. 2.3.

Personnel is usually not exposed to these hazards during operation, because in these phases, access to accelerator buildings or tunnels is forbidden or restricted. However, the life cycle of accelerator magnets includes test phases, for example after fabrication, for diagnostic purposes or after maintenance and repair. These tests are conducted in accessible locations: in laboratories or workshops, or in the accelerator area outside of operation. Personnel may stay close to the magnet to perform measurements, or they follow other occupations in the vicinity of the tested magnet.

2.2.3.1   Magnetic Field Hazard

During magnetic testing, personnel may be exposed to the magnetic fringe field penetrating out of the magnet yoke. This is unavoidable, for example, when making electrical or magnetic measurements for which the magnet must be powered. For accelerator dipole magnets, a rule of thumb is that the dipole field falls over two gap-widths by a factor of 5 [15]. In the example of the SPS dipole magnet (Table 2.1), the fringe field in 10 cm from the pole face would amount to 400 mT, the magnetic field gradient in these first 10 cm would be 16 T/m. There is presently no evidence of adverse health effects from an occasional exposure to magnetic fields of this magnitude.

2.2.3.2 Health Effects of Magnetic Fields

The International Commission on Non-Ionising Radiation Protection (ICNIRP) [12] has the aim to protect workers and members of the public from harmful effects of static and dynamic electromagnetic fields and of non-ionising radiation (electromagnetic radiation with wavelengths greater than 100 nm [12]).

In 2009, ICNIRP published exposure limits to static magnetic fields [13]. They investigated three interaction mechanisms between magnetic fields and living tissues: magnetic induction, magneto-mechanical and electrical interactions. A thorough analysis of the available research literature showed:

- No pronounced physiological effects have been found from exposure to fields of up to 8 T, except a small increase in systolic blood pressure.
- No evidence of health effects of exposures of up to 8 T on other aspects of cardiovascular function, on body temperature, memory, speech, or auditory-motor reaction time and of any other serious health effects in human volunteers.
- Magnetic fields of 2–3 T can cause transitory sensory effects including nausea, vertigo, metallic taste and phosphenes (light sensations induced in the retina and the optical nerve) when moving the head.
- The few available epidemiological studies of workers in aluminium smelters, chloralkaline plants or as welders do not indicate strong effects of exposure of up to several tens of mT on cancer incidence and reproductive health.

Based on this evidence, ICNIRP recommends that occupational exposure of the head and trunk shall not exceed a spatial peak magnetic flux density of 2 T. Exceptionally, for specific applications and in controlled situations, exposure of up to 8 T can be permitted.

Members of the public shall not be exposed to more than 400 mT on any part of the body.

The electrical circuits of implanted medical devices (pacemakers, cardiac defibrillators, insulin pumps etc.) can be perturbed by external magnetic fields, leading potentially to health risks for the implanted persons. Studies have shown that magnetic flux densities of less than 0.5 mT have no adverse effects on implanted medical devices and on implants from ferromagnetic metals. This leads to the practical requirement to put warning signs at the limit of areas where the magnetic flux density may exceed 0.5 mT.

2.2.3.3 Magnetic Forces

While there is no evidence for biological effects of fringe fields from accelerator magnets, a more imminent danger results from the mechanical forces exerted by the magnetic field on metallic objects.

An external magnetic field exerts a force on the magnetic moment of a metallic object. The magnetic moment can be induced, as in diamagnetic and paramagnetic materials or permanent, as in ferromagnetic materials. The induced magnetic moment of a metallic object with volume V is related to the external magnetic flux density by

$$\vec{m} = \frac{\chi}{\mu_0} \vec{B}_e V.$$

χ is the magnetic susceptibility, for ferromagnetic materials it can take on considerable values, which expresses the amplification of the external field within the ferromagnet. The force on the magnetic moment in an external field B_e is:

$$\vec{F} = \nabla \left(\vec{m} \cdot \vec{B}_e \right)$$

As an example, consider the force on a stainless-steel spanner ($\chi = 100$) in the inhomogeneous field close to the pole face of a dipole magnet. In an external field of $B = 1$ T, the induced magnetic moment in the spanner ($V = 200 \cdot 10 \cdot 3$ mm$^3 = 6 \cdot 10^{-6}$m^3) is 477 A m^2. The force on the spanner's magnetic moment in an inhomogeneous magnetic field with gradient of 1 T/m is of the order of 500 N. The likely result is that the tool would be wrought out of the hands of the worker and stick to the pole face of the magnet. This calculation is approximate, neither the susceptibility χ is precise, it depends on material composition and thermal history, nor is the inhomogeneous magnetic field. It only demonstrates the order of magnitude of the expected forces. At a larger distance, the magnetic force on a metallic object would be smaller, but often large enough to mobilise it and convert it into a projectile flying in the direction of the magnet poles.

The examples and conclusions above are valid for workers on powered accelerator magnets. The situation is different for workers in the vicinity or *inside* the large, superconducting solenoid magnets employed in particle detectors at high-energy accelerators. There, magnetic flux densities may attain up to 4 T, and access should only be permitted to trained volunteers, following specific procedures to minimize the sensory effects described in the literature.

As a summary of the section on magnetic field hazards one can observe

- Personnel is exposed to the fields from accelerator magnets only in exceptional situations, for example in test laboratories. The fringe field of dipole and higher order multipole magnets decays rapidly outside of the yoke and coils.
- In this configuration, physiological effects can be excluded. They would be expected at magnetic flux densities exceeding 2 T. Long-term health effects have not been observed in groups which are regularly exposed to fields of up to 100 mT, like welders.
- A limit of 0.5 mT is considered safe for persons with implanted medical electronic devices.
- Finally, the mechanical forces on ferromagnetic, metallic objects can be considerable and can project them in the direction of the magnet. To avoid injury, such objects must be banned from magnetic test stands.

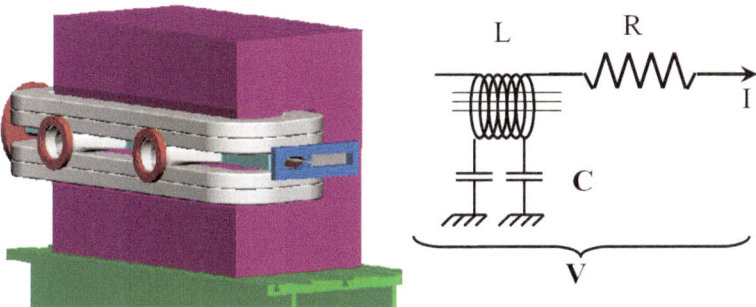

Fig. 2.8 A dipole bending magnet with its equivalent circuit [10]. Instead of U, the author of the figure uses V as symbol for voltage

2.2.3.4   Electrical Hazards of Magnets

Accelerator magnets are electrically powered. To generate magnetic fields in accelerators one needs generally high current ($I > 1$ A) and in some cases high voltage ($U > 1$ kV). The potential danger from a powered accelerator magnet shall be illustrated at the example of the SPS magnets. Figure 2.8 shows an equivalent electrical scheme of an electromagnet, composed of an ideal inductivity L, a resistance R (the ohmic resistance of the coil and cables), and a stray capacity to ground C.

The following relations apply:

magnet voltage:

$$U = RI + L\frac{dI}{dt};$$

instantaneous power:

$$P = UI = RI^2 + LI\frac{dI}{dt};$$

stored energy:

$$E = \frac{1}{2}LI^2 \cdot \frac{dE}{dt} = LI\frac{dI}{dt};$$

therefore power:

$$P = UI = RI^2 + \frac{dE}{dt}.$$

The dipole magnet of the CERN Super-Proton Synchrotron (SPS) (Table 2.2), powered by a continuous current of value I_{peak} for test purposes, stores an electromagnetic energy of $W = 163$ kJ. If one creates accidentally a short circuit between the terminals of this magnet, the circuit will drain the electrical power $P = UI$ over the resistance of the short circuit R_{sc} with a characteristic time constant of $\tau = \dfrac{L}{R_{sc}}$, which is in the order of a second. If this energy were dissipated entirely in the piece of metal causing the short, it would be enough to melt approximately 100 g of iron, giving an impression of the injury and material damage which can be created by an accidental electrical short circuit.

In the SPS, 784 dipole magnets of two types with slightly different parameters are connected in series. Resistance and inductance of the magnets add up to values of $R_{tot} = 3.25\ \Omega$ and $L_{tot} = 6.6$ H. The stored energy of the full SPS dipole circuit amounts to 109 MJ. At peak current, the resistive voltage across R is 18 kV, to which the induced voltage over L_{tot} must be added. The SPS magnets are ramped at 1.9 kA/s, adding the induced voltage $V_L = L\dfrac{dI}{dt} = 12.5\,\text{kV}$. Therefore, the insulation of the magnets must withstand 30 kV.

If the powering circuit of the magnet is suddenly opened, the induced voltage

$$U_i = -L\frac{dI}{dt}$$

appears across the interrupter gap. The time derivative of the current is initially very high, and the high inductive voltage will result in an electrical arc across the gap. This electrical arc carries high electrical power, and its primary hazard is heat radiation and projection of fused metal, not electrical shock. More on arc flashes is found in Sect. 4.1.1.

2.3 Cryogenics

With *cryogenics* one designates the realms of science and technology operating at temperatures lower than 120 K, where gases such as nitrogen, oxygen or argon begin to liquify [27]. Hazards of cryogenics are related to very low temperatures and to oxygen deficiency (Fig. 2.9).

Cryogenic technology is essential for the function of a modern high-energy particle accelerator using superconducting magnets and radiofrequency cavities. Superconducting magnets are also used in Nuclear Magnet Resonance (NMR)

Fig. 2.9 Warning signs against low temperatures and oxygen deficiency hazard, after [1]. (Image source: https:// publicdomainvectors.org)

imaging in medical diagnostics, and cryogenics has become a key technology with a certain economic impact.

The purpose of accelerator cryogenics is to cool magnets and RF cavities to temperatures well below the critical temperature for superconductivity T_c (see 2.2.2) to ascertain their functioning with enough operational margin. The cool-down is achieved in a heat-exchange process with helium as a cryogenic fluid, the only element apart from the explosive hydrogen which does not become solid other than under extreme pressures.

Due to the size of a modern high-energy accelerator, the cryogenic installations have industrial dimensions. The cryogenic plants at CERN are world-wide the largest of their kind.

2.3.1 Production of Low Temperatures

The physical principle of cooling an object is to transfer its thermal energy to a reservoir at a higher temperature by a heat exchange mechanism using mechanical work. The cryogenic fluid which is produced in a refrigeration plant can be transported to the object to be cooled where it absorbs part of the object's thermal energy. In this process, the fluid may undergo a phase change from liquid to gaseous. Finally, the "warm" fluid is transported back to the refrigeration plant from where on the cycle repeats. This is the working principle of the refrigerator or the heat-pump.

The refrigeration plant makes use of a thermodynamic process to extract heat from the cryogenic fluid (Fig. 2.10). Heat is extracted from a low-temperature

Fig. 2.10 Open thermodynamic cycle, transporting thermal energy from low temperature (T_c) to high temperature (T_w) by employing mechanical work W. (From [24])

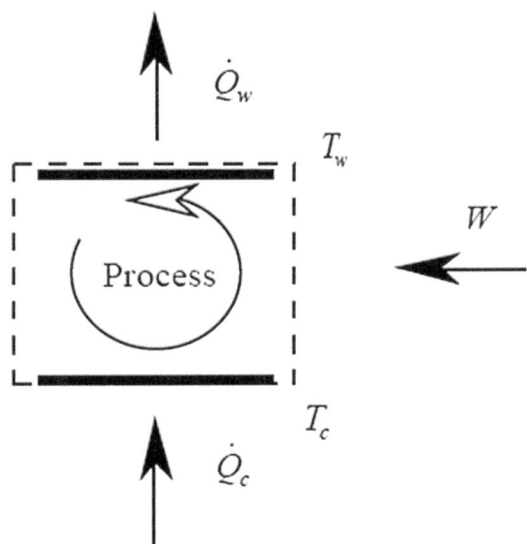

reservoir (T_c) and mechanical energy W is used to bring it to a higher temperature T_w where it is rejected to the environment. This is the principle of the heat pump, but in cryogenics, the interest is not the heat flow at the high temperature $\dot{Q}_w$ but the heat flow extracted from the cold medium Q_c , in order to lower its temperature further. The cyclic thermodynamical process with the highest efficiency is named after the French physicist Carnot [18]. Accordingly, any thermodynamical process to extract heat energy from a cold reservoir will need at least a mechanical energy W

$$W > \dot{Q}_c \left(\frac{T_w}{T_c} - 1 \right)$$

The expression in parenthesis is called the *Carnot factor*".

It indicates that even in an ideal Carnot cycle one must inject $\left(\frac{300}{4} - 1 \right) = 74$ W of process energy at room temperature (300 K) to remove 1 W of thermal energy at the cryogenic temperature of 4 K.

The Carnot cycle can only be approximated, and numerous technical cycles have been developed for application in real refrigeration plants, see for example [26]. All thermodynamic cycles used for cryogenics have in common that the temperature of the gas is reduced by first charging it isothermally with mechanical energy in a compression step and then letting the gas expand, for example in a nozzle (Joule-Thompson process) or more efficiently, by letting it perform work against the resistance of a turbine.

Due to the unfavourably low thermodynamic efficiency, the refrigeration plants for a particle accelerator have industrial dimensions. CERN is presently operating the world's largest helium refrigeration plant, distributed over 8 sites, to cool the approximately 20 km of superconducting magnets in the LHC to a temperature of 1.9 K. These plants have a high engineering complexity and they are usually conceived and constructed in collaboration with specialised engineering firms.

Safety aspects concerning the construction of the compressors and refrigerators are taken account of by their manufacturer who constructs them following industrial standards. In the European Union, the mandatory application of the Machinery Directive (see 4.2.2) guarantees minimal health and safety standards across the member states and for imported products.

The noise level inside of a cryogenic service building can be overpowering and easily exceed the legal limits (Sect. 4.5). The manufacturers of compressors and turbines take all possible steps to reduce the emissions form their devices. Further attenuation of compressor noise is often impossible because additional insulating material would impede the heat exchange of the plant with its environment and lead to overheating. Cryogenic compressors and refrigerators are installed in specific buildings with isolated walls to reduce noise emission to the environment. The control rooms for these plants are preferably installed in a separate building, or at least isolated so well that the sound pressure level does not exceed a comfortable level.

Table 2.3 Properties of cryogenic fluids used in accelerators and particle detectors. M molar mass, T_b boiling temperature at 1013 hPa, ρ_l liquid density at T_b, ρ_g gaseous density at T_b, ρ_{STP} gaseous density at standard pressure and temperature (1013 hPa and 0° C), ΔH_V enthalpy of evaporation. After [20]

Fluid	M (g mol^{-1})	T_b (K)	ρ_l (kg m^{-3})	ρ_g (kg m^{-3})	ρ_{STP} (kg m^{-3})	$\rho_l/\rho_{STP} = V_{STP}/V_L$	ΔH_V (kJ kg^{-1})
Helium	4.003	4.2	124.9	16.9	0.178	702	20.3
Hydrogen (H$_2$)	2.016	20.3	70.8	1.34	0.0899	788	446
Nitrogen (N$_2$)	28.01	77.3	808	4.62	1.25	646	199
Argon	39.95	87.3	1395	5.77	1.79	780	163

2.3.2   Cryogenic Fluids

The most common cryogenic fluid used in particle accelerators is Helium. It has a low enough boiling temperature to remain liquid in the temperature range where most low-temperature superconductors operate. Another cryogen of technical importance is Nitrogen, whereas liquid Argon is used in some particle detectors. In the past, Hydrogen was used in bubble chambers, particle detectors which have been replaced by modern alternatives. Table 2.3 lists some physical properties of these fluids.

Cryogenic fluids have in common

- a very low temperature in the liquid state, and
- a high ratio of gaseous volume at ambient temperature to liquid volume $\rho_l/\rho_{STP} = (650–800)$.

Because of its low enthalpy of evaporation, Helium must be thermally isolated from all heat transfer by conduction, convection, or radiation to remain in the liquid state. This can be achieved with *super-insulated cryostats,* combining vacuum insulation to interrupt convection, metallised mylar foils as radiation barrier and low-conduction mechanical supports from organic materials. If the insulation of a cryostat fails, two effects are the consequence:

- the heat flow in the cryogen will lead to its partial evaporation, increasing rapidly the internal pressure in the cryostat (overpressure),
- once the cryogenic fluid is released from the cryostat, persons may be exposed to cold and to oxygen deficiency.

2.3.2.1   Overpressure

In case the thermal insulation of a cryostat fails, for example by a degradation of the insulation vacuum, thermal energy is transported from the surroundings to the cryogenic vessel by a combination of conduction and convection. A part of the cryogenic fluid evaporates, and by the relation for an ideal gas $pV = $ const., the pressure in the fixed-volume vessel increases. From Table 2.3, the conversion of liquid helium to gaseous helium at the boiling temperature leads to a pressure increase by a factor of 7.4. Further heating of the gas leads to higher pressure. Obviously, cryostats constitute *pressure vessels*, containers built and tested to withstand pressures higher than atmospheric pressure. The topic of pressure vessels is treated in more detail in Sect. 4.2. To avoid damage to the cryostat from the increasing internal pressure, they are equipped with so-called safety devices, in form of pressure relief valves. An international standard [22] describes the models and calculations to determine the correct dimension of the safety valves on cryogenic pressure vessels. For cryogenic liquids, these calculations are far from trivial because often the cryogenic fluid undergoes a phase change from liquid to gas during release, or a mixture of liquid and gaseous phases of the fluid is released.

2.3.2.2   Cold Burns

The exposure of skin to cold surfaces or cold fluids may lead to "cold burns", named after the similarities of their effects on tissue to those by heat or flame. When exposed to temperatures well below 0 °C for longer than the temperature regulation mechanism of the body can cope, the skin develops in a first phase reddening and blisters, which may be painful but reversible (1st degree). The cold may penetrate layers below the skin and lead to tissue damage by the formation of ice crystals, leading to irreversible effects, like tissue necrosis. These effects are well-known to high-altitude mountaineers and arctic explorers but may as well occur in the workplace when exposed to cryogenic liquids.

To prevent cold burns, one must prevent the exposure to extremely cold surfaces or fluids. Protection measures are:

- Orientation of cold gas relief devices (safety valves) away from passageways.
- Insulation of cryogenic vessels and pipes, so that no cold spots appear on the external surface.
- Wearing correct personal protective equipment when working on cold cryogenic systems. This consists of long-sleeved trousers and jackets, temperature-insulating gloves which remain flexible at low temperatures and a face shield or at least safety goggles.
- Never walk across the condensation cloud around a release of cryogenic fluids. The lack of visibility makes it impossible to avoid the jet of cold liquid and severe injury might result.

2.3.3 Oxygen Deficiency Hazard from Cryogenic Fluids

Everybody can feel the effects of a reduced oxygen concentration in air during a visit in the high mountains. After leaving a cable car at an altitude of nearly 3900 m above sea level[2] one may feel dizziness and headache. During longer stays at high altitude, for example when climbing a summit or staying overnight in a mountain refuge, tiredness and vomiting may add to the symptoms of what is called altitude sickness. This is the mildest form of the consequences of hypoxia or oxygen deficiency, the term used in the occupational safety context. It is caused by the lower partial pressure of oxygen, which is 21% at sea level, but only 13% at 3900 m altitude. The physiological effects of reduced oxygen concentration are listed in Table 2.4

Cryogenic fluids experience a volume expansion by a factor of 700 to 800 upon evaporation, and they can displace or dilute breathable air, creating an atmosphere depleted of oxygen. Therefore, a risk assessment must evaluate the likelihood and consequences of oxygen deficiency at workplaces where cryogenic liquids are used.

2.3.3.1 Oxygen Deficiency from Nitrogen Gas

Liquid Nitrogen (LN_2) is employed in cryogenic shields, or to cool high-temperature superconductors. Once gaseous, it mixes perfectly with air of which it is the major constituent.

A first, coarse estimate of oxygen deficiency hazard in a closed room with volume V_o is obtained by assuming that the evaporated volume of the released LN_2, $V_{R.g}$, displaces an equivalent volume of air from the room and that the remaining air and cryogen mix perfectly (Fig. 2.11). In this case, the oxygen concentration of the mixture is:

Table 2.4 Physiological effect of reduced oxygen concentration [21]

O_2 (Vol %)	Effects and Symptoms
18–21%	No discernible symptoms can be detected by the individual.
	A risk assessment must be undertaken to understand the causes and determine whether it is safe to continue working.
11–18%	Reduction of physical and intellectual performance without the sufferer being aware.
8–11%	Possibility of fainting within a few minutes without prior warning.
	Risk of death below 11%
6–8%	Fainting occurs after a short time.
	Resuscitation possible if carried out immediately.
0–6%	Fainting almost immediate. Brain damage, even if rescued

[2]In the Alps, this is possible on the summits of Aiguille du Midi (Chamonix, FR) and Klein Matterhorn (Zermatt, CH).

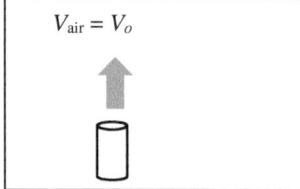 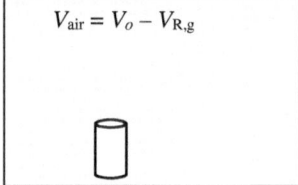

Fig. 2.11 Simplest model to assess oxygen deficiency in a room. Left panel: (liquefied) gas is streaming into a room with volume Vo. Right panel: the expanded cryogenic liquid or gas has displaced an equal volume of air. The oxygen concentration is calculated form the remaining air volume. (Copyright: the author)

Fig. 2.12 Substance and air flows entering and leaving the room (or the control volume) in the well-mixed room approximation. (Copyright: the author)

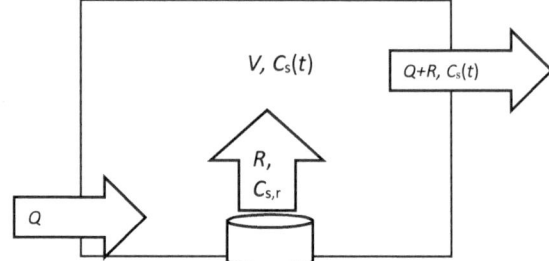

$$C\left(O_2\right) = 0.21 \cdot \left(1 - \frac{V_{R,g}}{V_o}\right)$$

If the oxygen concentration shall not drop below 18%, the lowest concentration which is universally regarded as safe for workers, then the volume of the room must be more than 7 times the volume of the evaporated gas:

$$V_0 > 7 V_{R,g} \gtrsim 5000 V_{R,l}$$

This relation makes use of the typical volume ratio between the cryogenics gaseous state gas at 300 K and its liquid volume. This can be expressed as a rule of thumb: *"No oxygen deficiency hazard in a room with volume in m³ exceeding the cryogenic liquid volume in L by at least a factor of five"*.

A more realistic estimate can be obtained from the *well-mixed room approximation*. Figure 2.12 shows the schematic of a room into which a substance S is released with a volume flow R (m³ s⁻¹) and with a relative concentration $C_{s,r}$ (Volume-% or m³ m⁻³). Fresh air is supplied with a ventilation flow Q (m³ s⁻¹). It is assumed that pressure in the room remains constant. Therefore, no accumulation of air or gas can happen, and the outflow of the room is $(Q + R)$.

The time dependent concentration of substance S in the room is denoted $C_s(t)$. The rate equation for $C_s(t)$ is:

$$V\frac{dC_s(t)}{dt} = RC_{s,r} - (Q+R)C_s(t)$$

If the release of substance S stops after time T, then the solution for the time-dependent concentration of s, $C_s(t)$ is during the release:

$$t \leq T : C_s(t) = \frac{RC_r}{(Q+R)}\left[1 - \exp\left(-\frac{Q+R}{V}t\right)\right].$$

After the release, the concentration of S develops as:

$$t > T : C_s(t) = C_s(T)\exp\left[-\frac{Q}{V}(t-T)\right]$$

$$= \frac{RC_r}{(Q+R)}\left[1 - \exp\left(-\frac{Q+R}{V}T\right)\right]\exp\left[-\frac{Q}{V}(t-T)\right].$$

During a release of LN_2, the concentration of the released gas (the evaporated liquid) is 100%. One is interested in the remaining concentration of oxygen in air, which is $C_{O_2}(t) = 0.21(1 - C_s(t))$. It follows for times during the release:

$$C_{O2}(t) = 0.21\left\{1 - \frac{R}{(Q+R)}\left[1 - \exp\left(-\frac{Q+R}{V}t\right)\right]\right\}$$

$$= \left(\frac{0.21}{Q+R}\right)\left\{Q + R\exp\left(-\frac{Q+R}{V}t\right)\right\}$$

For long times, the equilibrium concentration $C_{O2,eq}$ is attained:

$$C_{O_{2,eq}} = 0.21\left(\frac{Q}{Q+R}\right)$$

If a minimum concentration of 18% O_2 shall be preserved, one can evaluate the minimal ventilation flow Q_{min} required to cope with the worst-case release rate R_{max} of the cryogenic gas:

$$\frac{0.18}{0.21} = \frac{Q_{min}}{Q_{min} + R_{max}}$$

This equation can be solved for the result $Q_{min} = 6\,R_{max}$.

A numerical example shall illustrate this relation: a liquid nitrogen cooled circuit loses 10 mL s^{-1} LN_2, evaporating to 6.5 L s^{-1} N_2-gas at STP. To retain an oxygen concentration above 18%, the ventilation must supply 140 m^3 fresh air per hour. In a standard room one considers that air is exchanged between 2 and 3 times per hour. This means that the assumed leak rate of LN_2 of 10 mL s^{-1} does not pose a problem in normally sized laboratories. It could however lead to oxygen deficiency in a small wall cabinet in which a leaking LN_2-dewar is stored.

The well-mixed room approximation gives more reasonable estimates than the simple displacement hypothesis if the released gas is indeed well mixed. In not too large rooms, the turbulence created during the outflow and evaporation of cryogenic liquid suffice to promote mixing in the room. Refinements of the well-mixed room approximation can be found in [23]. A strong ventilation system supplying fresh air provides an effective protection against local pockets of oxygen deficiency in a room even for stronger releases that in the example above.

2.3.3.2 Oxygen Deficiency from Helium Gas

The most important cryogenic fluid for superconducting magnets is liquid Helium (LHe). The oxygen deficiency hazard which may arise after a release of LHe cannot be described with the well-mixed room approximation because of its low density once in gaseous state. Below a temperature of 40 K the density of Helium gas exceeds the density of air, 1.2 kg m^{-3}. The heat capacity of Helium is very low and little energy is required to warm it up. Thus, liquid Helium or cold He gas released to air rapidly warms up by radiation and by turbulent convection. The gas density diminishes proportionally to 1/T. Once the temperature of the outpouring gas exceeds 40 K, buoyancy drives the gas upwards. In closed rooms the warming-up Helium gas forms a layer under the ceiling. A quick estimate yields that the thickness of the helium layer in a room with surface area A attains 5.6/A m for each kg of Helium released. In a standard laboratory room with a height of 3 m and a surface of 4 m by 7 m, 7.5 kg Helium (60 litres of LHe) would displace half the air from the room and form a layer from the ceiling extending until a height of 1.5 metres.

In laboratories housing experimental Helium cryostats, simple preventive measures can be taken: exhaust fans placed at a small distance under the ceiling expel the helium gas from the room, at the same time a vent at floor level supplies fresh air.

To evaluate the oxygen deficiency hazard in accelerator tunnels, a series of Helium release experiments have been performed in CERN's LHC [19, 25]. 1000 litres (125 kg) of liquid Helium were released into the LHC tunnel with three different mass flows: 100 g/s, 320 g/s and 1 kg/s. The tunnel ventilation worked at its standard speed of about 1.5 m/s. Downwind from the release point, oxygen concentration sensors were located at regular height and distance in the tunnel and the propagating Helium cloud was recorded at these points by video cameras. The results of these tests can be summarized as follows:

- For releases of mass flows with 100 g/s or 320 g/s, a turbulent zone with an extension of a few metres around the release point develops, in which the released cold helium is vigorously mixed with the surrounding air. In this zone, the Helium warms up rapidly and raises by buoyancy to the ceiling. In some distance from the release point, a ceiling flow of helium is driven by the ventilation and by its own momentum towards the next shaft. Outside the turbulent zone, temperature, and oxygen levels at the height of a walking person ($h < 2.5$ m) are close to normal and permit a safe evacuation of personnel.

- A release of Helium with a mass flow of 1 kg/s corresponds to a gaseous Helium volume of 5.6 m^3/s at standard pressure and room temperature. The thermalized Helium gas displaces air in most of the tunnel cross section so that oxygen is rarefied over several hundred metres.

The Helium release experiments have yielded precious information for the operation of the LHC and for the safe design of future accelerator facilities. Few publications of theoretical/ numerical descriptions of cryogenic releases exist [25] and in general, this field is actively researched.

As protection measures it has been decided that personnel are not permitted to access the LHC tunnel in operational phases, where leakages may provoke helium releases of more than 320 g/s. This is the case during the powering of the superconducting magnets, when also electrical safety demands the exclusion of personnel form the tunnel area.

In periods when the magnet current is zero, a general risk assessment of the cryogenic system has shown that accidentally provoked helium releases cannot have a mass flow exceeding 100 g/s. In these phases, personnel are permitted to perform standard verifications and small maintenance after a specific risk assessment. This assessment is targeted to identify the risk of an accidental manipulation or damage of a cryogenic control device, potentially leading to a Helium release. Technical or organisational mitigation measures are prescribed in the risk assessment to reduce this probability.

2.3.3.3   Mitigation of Oxygen Deficiency Hazard

The following mitigation measures can be applied to reduce the risk that personnel are exposed to the oxygen deficiency hazard:

- Process changes: consider storing the bulk of cryogenic fluid outside of regularly occupied rooms and introduce only the immediately required quantity.
- Forced ventilation: in unventilated or poorly ventilated rooms, install a forced ventilation with sufficient flow to remove enough of the evaporated cryogenic liquid to remain at $C(O_2) > 18\%$
- Oxygen deficiency detectors: they determine $C(O_2)$ by an electrochemical reaction and raise an alarm when $C(O_2)$ decreases below a set point. In case of ODH alarm, personnel must leave the room immediately. Due to their larger volume, installed ODH detectors are more precise, on the other hand, portable personal ODH monitors are light-weight and less expensive, but less precise and prone to raise false alarms.

Fig. 2.13 Warning sign
against electromagnetic
fields after [1]. (Image
source: https://
publicdomainvectors.org)

If neither of the technical measures above can sufficiently reduce the risk of exposure to ODH, then the workers accessing the hazard areas must carry personal protective equipment in the form of personal oxygen generators. Adopted from underground mining, these devices supply O_2 from a chemical reaction for approximately 30 min. They enable a safe evacuation in large facilities, such as accelerator tunnels, where the emergency exit may be far.

2.4 Radiofrequency Technologies

Radiofrequency technology is fundamental to the acceleration of particles (Fig. 2.13). An alternative, emerging technology is plasma wake field acceleration, which will be briefly touched in Sect. 2.5.1.

2.4.1 Principle of RF Acceleration

A charged particle can be accelerated by static voltage across a gap, but this simple acceleration principle is limited by the breakdown voltage of the electrical field in vacuum to a maximal voltage of a few MV. The fundamental idea of radiofrequency acceleration is to apply an alternating electrical field to a gap. The electrical field has the accelerating polarity when the charged particle passes the gap, and the particle is shielded from the decelerating polarity, for example in drift tubes in a linear accelerator, or simply by having left the gap in a circular accelerator. Figure 2.14 illustrates this principle.

A particle moves in z-direction. If the electrical field was a constant across the gap g:

$$E_z = E_{z,0} = {V_0}\!\big/\!{g} \, ,$$

A particle with electrical charge e would gain the kinetic energy $\Delta E = e\, V_0$. In reality, the electrical field oscillates with frequency ω:

Fig. 2.14 Schematic illustration of a single accelerating gap. (From [28])

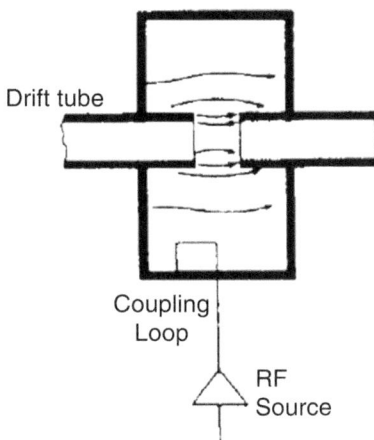

Drift tube

Coupling Loop

RF Source

$$E_z(t) = \frac{V_0}{g} \cos(\omega t)$$

The field is at its maximum when the accelerated particle is in the centre of the gap. Then, the energy absorbed from the electrical field is:

$$\Delta E = e V_0 T$$

where T is the transit-time factor. For the simple example of a sinusoidal field variation across a gap, $T = 0.637$. The energy gain of the charged particle per passage through the gap is determined by

- The maximum Voltage V_0
- The transit-time factor, which can be influenced by the geometry of the accelerating structure,

An accelerating gap is the simplest topology of an RF accelerating structure. Other topologies are drift-tube linacs (a succession of gaps, alternating with shielding drift-tubes) or accelerating cavities, in which standing or travelling electromagnetic waves impart kinetic energy on the particle.

2.4.2 Components of a RF Acceleration System

Figure 2.15 shows a block-diagram of an RF accelerating system. An amplifier converts DC input voltage into RF (high-frequency AC) output power. The power is transmitted by waveguides or coaxial cables to the accelerating cavity, where it is transferred to the beam particles.

2.4.2.1 RF Amplifier

An RF amplifier converts electrical power from a DC source into radiofrequency output power. In Fig. 2.15, the amplifier converts a DC input to radiofrequency output, modulated by the low-level RF signal at the amplifier's control input. Frequency and phase of the RF output signal are determined by the low-level RF power, its amplitude and power are proportional to the current and voltage of the DC input. In the past, RF amplifiers were based on some type of vacuum tube. Today, technical solutions with solid-state amplifiers are possible.

In tetrode vacuum-tube RF amplifiers for CERN's Super-Proton Synchrotron (SPS), packages of electrons were accelerated by anode voltages of 24 kV. In a Klystrons, a DC electron beam modulated by the low-level RF is accelerated up to 100 kV. When high-energy electrons collide with matter, they emit a Bremsstrahlung X-ray spectrum. In other words, vacuum tube-based RF amplifiers are sources of ionizing radiation

2.4.2.2 Waveguides

For operational reasons, the RF amplifiers are placed at a certain distance from the particle beam so that they can be monitored while the accelerator is working and serviced without having to enter a radiation area. The RF power from the amplifier is transported by waveguides or coaxial cables to the accelerating cavity in the accelerator tunnel. Above a frequency of a few 100 MHz and power exceeding some 10 kW, hollow waveguides are the most efficient means to transport RF energy. Electromagnetic waves can travel in hollow, metallic guides with rectangular cross section, when certain relations between the wavelength and the dimensions of the guide are met. These relations are derived in standard textbooks of electrodynamics, e.g [31]. If waveguides are incorrectly mounted, leakage radiation may escape. This radiation is non-ionising, electromagnetic radiation (NIR) and, depending on its power, may have adverse effects on persons in the vicinity.

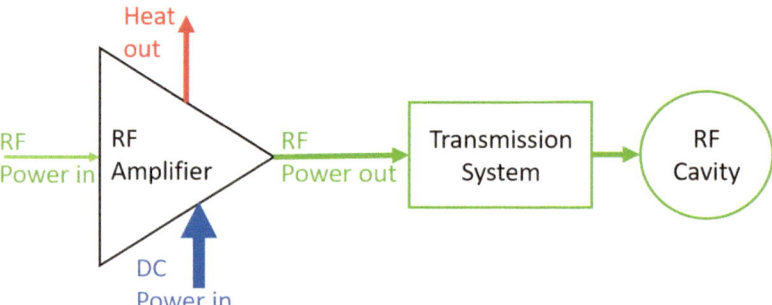

Fig. 2.15 Generic lay-out of an RF system, after [29]

2.4.2.3 RF Cavity

The shape of the metallic RF cavity is designed to maintain a standing or travelling radiofrequency electromagnetic field of the required frequency. The surface resistance of the metal constitutes a loss factor, and can be drastically reduced by making the cavity walls superconducting, for example by covering them with a thin layer of Nb.

2.4.3 Hazards from RF Systems

Radiofrequency systems may be the source of the following hazards which must be considered in a risk assessment:

- Electrical hazards from the power source (Sect. 4.1)
- Cryogenic hazards in superconducting RF systems (Sect. 2.3)
- X-rays from bremsstrahlung from the electron beam in vacuum tube RF amplifiers or from spuriously emitted and accelerated surface electrons
- Electromagnetic radiation leaking from radiofrequency systems, with frequencies too low to provoke ionisation: Non-Ionising Radiation.

2.4.3.1 Bremsstrahlung X-Rays

The X-ray bremsstrahlung spectrum from vacuum tube RF-amplifiers must be appropriately shielded. It has a terminal energy corresponding to the highest kinetic energy of the electrons in the vacuum tube, in some klystrons exceeding 100 kV. Exposure to ionizing radiation from RF amplifiers made the headlines in 2001 in Germany, when it was revealed that maintenance technicians of military RADAR devices in tanks and airplanes, fed by klystrons or magnetrons, showed a higher cancer incidence than the general population. This was traced back to their exposure to the partially unshielded radiation emitted by these devices during maintenance. Coincidentally, this occurred in both of the armed forces of former West- and East Germany. Vacuum tube RF amplifiers must be appropriately shielded to keep the emitted radiation below legal limits for non-designated areas. Klystrons bought from industry are delivered with shielding, and the radiation risk from them is low if the shielding is not compromised.

In accelerating cavities, electrons are emitted spuriously from the cavity surface at locations of high curvature, surface inhomogeneities or "dirt". These electrons pick up a certain amount of the RF energy in the cavity and collide with its walls, emitting bremsstrahlung. The intensity of the emitted X-rays is the highest, when a cavity recently exposed to air is *conditioned*, a process where surface electrons are emitted purposefully to contribute to the removal of impurities from the surfaces. When RF cavities operate in the accelerator tunnel, these emissions do no present a

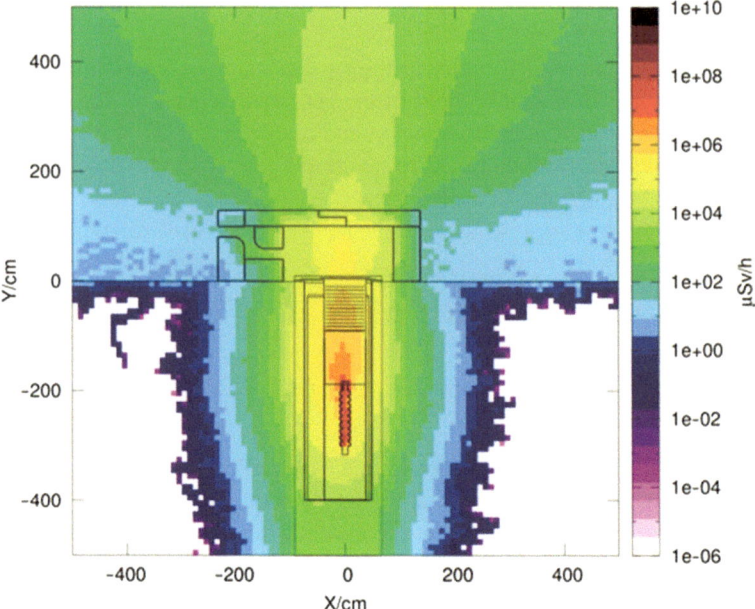

Fig. 2.16 Dose rate of ionizing radiation around a vertical test stand for RF accelerating cavities, from [31]. (Image reused with permission by Oxford University Press)

risk for personnel, protected by the accelerator shielding and excluded by the access safety system (Sect. 5.3.1). During commissioning of new or modified cavities in test areas, proper shielding must be built around the test stands to protect the personnel from harmful ionizing radiation. Figure 2.16 shows a simulated contour map of ambient dose equivalent rate around a test bunker during the commissioning of RF accelerating structures with a field gradient of 25 MV/m in Beijing University [31].

In a two-step process, first the emission and spurious acceleration of surface electrons was simulated, and then the generation of bremsstrahlung X-rays from the stopping electrons. The light blue colour on the dose rate map outside of the bunker signifies a dose rate between 10 and 50μSv/h. This value would lead to a classification as a controlled radiation area under most European radiation protection regulations.

2.4.3.2   Electromagnetic Leakage Radiation

Non-ionising electromagnetic radiation leaking from RF structures and wave guides may have negative effects on human health. The subject of the next Sect. 2.4.4 are direct health effects of electromagnetic field exposure of the whole human body or of parts of it.

Indirect health effects occur when electromagnetic radiation alters or impedes the function of electric devices. A heart pacemaker or an insulin pump are electrical or electro-mechanical devices, equipped with electronic control circuitry. Any of these components may fail if they are coming under the influence of electromagnetic fields. The avoidance of such adverse effects is the subject of electromagnetic compatibility (EMC).

2.4.4 Health Effects of Electromagnetic Fields (EMF)

The field of non-ionising radiation covers a wide band of frequencies of electromagnetic radiation. The quantum energy of its photons is not high enough to provoke the ionisation of an atom in material struck by the radiation, therefore its name *non-ionising radiation* as compared to ionising radiation.

Nevertheless, the effect of exposure to non-ionising electromagnetic radiation can be as detrimental to health as that of ionising radiation.

The International Commission for Non-Ionizing Radiation Protection, ICNIRP, has updated its recommendations for electromagnetic radiation in the frequency range between 100 kHz and 300 GHZ in 2020 [30]. ICNIRP bases its recommendations for intensity limits of electromagnetic radiation only on substantiated effects. These are harmful effects of EMF on health which have been repetitively observed and documented in the scientific literature and which do not contradict current scientific understanding. The rigorous approach to rely on scientific evidence for adverse health effects of EMF ensures that ICNIRP recommendations do not follow unconfirmed claims.

2.4.4.1 Biophysical Effects of EMF

The biophysical effects of electromagnetic fields in the body can be classified in three groups:

- The most important effects of EMF in the body are thermal effects. The electrical field components of EMF act on polar molecules (water!) and on charge carriers (ions and electrons) and confers kinetic energy to their translational, rotational, and vibrational degrees of freedom. This kinetic energy dissipates in the tissue, leading to a temperature rise. The adverse health effects of a continuous rise of body temperature have been demonstrated, and intensity limitations of non-ionising radiation exposure protect against these effects.
- At frequencies around $f \approx 100$ kHz, nerve stimulation which manifests itself in a "tingling" sensation has been described.
- Very high intensities of high-frequency EMF, which may occur in the Fourier-decomposition of electromagnetic pulses, may lead to an alteration of the permeability of biological membranes. Such modifications in the body would lead to

consequential adverse health effects. The effect has been demonstrated in in-vitro experiments, but at an intensity which is prevented by the limits on temperature effects.

As a summary, one can state that as of today, ICNIRP recognizes as the only substantiated adverse effect of non-ionising radiation on human health and safety the heating of exposed tissue. Spurious reports of electro sensitivity or the induction of cancer have not withstood epidemiological investigations.

2.4.4.2 Dosimetric Quantities for Non-ionising Radiation

To describe the biophysical effects from EMF leading to adverse health effects and to define limits, one needs dosimetric quantities. ICNIRP defines quantities for energy absorption from EMF in its report [30]. The quantities which are suitable for limiting emissions from RF fields at particle accelerators can be described as follows:

- *SAR,* Specific energy absorption rate, measured as a whole-body average in the unit $W\ kg^{-1}$, to quantify whole body energy absorption rate, leading to a rise of body core temperature
- SAR_{10g}, Specific *local* energy absorption rate over a 10 g tissue cube (with side length of 2.15 cm) to quantify local energy absorption rate, leading to local heating
- S_{ab} Specific Absorbed energy density, measured in the unit $W\ m^{-2}$, to quantify energy absorption rate to the skin from very high frequency EMF

2.4.5   Protection against NIR

2.4.5.1 Effect Levels and Basic Restrictions to Temperature Rise

When discussing body temperature rise, one distinguishes *core temperature* and *local temperature*. The *core temperature* is the temperature of the inner organs and amounts for a healthy adult to $T_{core} = 37\ °C$. Core temperature may rise in immune reactions (fever) to help the body fend off infections, but in absence of fever, a long-term temperature rise of more than $\Delta T_{core} = 1\ °C$ can lead to adverse health effects. Locally, body tissues shall not be heated by more than 5 °C to prevent effects, such as burns.

ICNIRP defines an effect level, the magnitude of an EMF, expressed in a suitable dosimetric quantity, from which on health effects are reported in the literature.

Modelling the effect of EMF on body tissue shows that a specific energy absorption rate $SAR = 4\ W\ kg^{-1}$ over 30 minutes raises body core temperature by 1 °C. This level of SAR is therefore the effect level for whole-body exposure. As a comparison, an average human generates $1\ W\ kg^{-1}$ at rest, $2\ W\ kg^{-1}$ standing and $12\ W\ kg^{-1}$

when running. This heat loads are dealt with by the body's temperature regulation system. Excess heat is transported to the skin, cooled by the surrounding air and, when this is not sufficient, by evaporation of sweat. The ICNIRP places a basic restriction on *additional* heat loads, which may overload the body's regulation mechanism.

If only part of the body is exposed to EMF, then local effect levels for energy absorption rate apply. They are determined by the quantity SAR_{10g}. The effect levels are $SAR_{10g} < 20$ W kg^{-1} for head and torso and $SAR_{10g} < 40$ W kg^{-1} for the limbs. As tissue has an assumed density of $\rho = 1$ g cm^{-3}, the local power density must not exceed 20 mW cm^{-3} or 40 mW cm^{-3}, respectively.

At frequencies above 6 GHz, the heating by EMF takes place mostly in the skin: at 6GHz, 86% of the power is absorbed within 8 mm of the body surface and leads to a rise of the *local temperature*. The suitable dosimetric quantity to describe energy absorption in this frequency range is specific absorbed energy density, S_{ab}. Here, effects over a small (1 cm^2) and a larger (4 cm^2) patch of skin are investigated, leading to two different effect levels of 400 W m^{-2} and 200 W m^{-2}.

To derive basic restrictions to exposure by EMF for workers, ICNIRP divides the effect levels by a safety factor of 2 or 10 (for whole-body *SAR*). For the public, basic restriction levels are reduced by another factor of 5 (Table 2.5).

2.4.5.2 Reference Quantities for Non-ionising Radiation

Basic restrictions are defined within the body or parts of it and cannot be measured. For the assessment of non-ionising radiation at the workplace or in the public, reference quantities are defined which are accessible to measurement. Reference

Table 2.5 Basic restrictions to the effects of electromagnetic radiation [30]. The table shows the body part to be protected, the frequency range, the maximal allowed temperature rise over a distinct volume, the applicable dosimetric quantity, the level, of the quantity at which health effects are observed with certainty and the derived exposure limits for workers and the public. The thermal effect is averaged over 6 min, with exception of the one on the whole body where the averaging interval is 30 min

Body part	Frequency	ΔT in volume	Quantity	Health effect level	Limit (Workers)	Limit (Public)
Whole body	100 kHz–300 GHz	1 °C.	*SAR*	4 W kg^{-1}	0.4 W kg^{-1}	80 mW kg^{-1}
Head & Torso	100 kHz–6 GHz	2 °C e	SAR_{10g}	20 W kg^{-1}	10 W kg^{-1}	2 W kg^{-1}
Limbs	100 kHz–6 GHz	5 °C	SAR_{10g}	40 W kg^{-1}	20 W kg^{-1}	4 W kg^{-1}
Skin	6 GHz–300 GHz	5 °C over 4 cm^2	S_{ab}	200 W m^{-2}	100 W m^{-2}	20 W m^{-2}
		5 °C over 1 cm^2		400 W m^{-2}	200 W m^{-2}	40 W m^{-2}

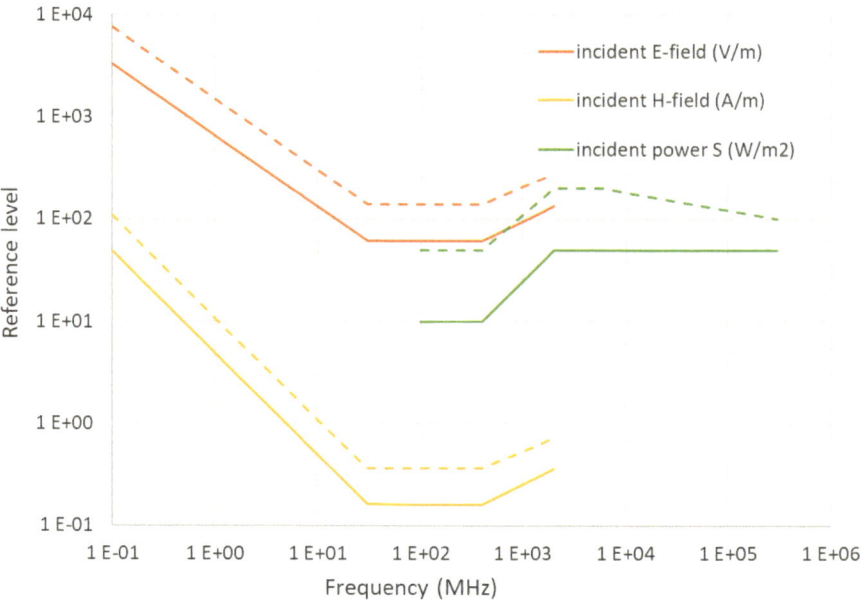

Fig. 2.17 Reference levels for occupational exposure to RF electromagnetic fields. Continuous lines: whole-body exposure, dashed lines: local exposure. (Data from Tables 5 and 6 in [30])

Fig. 2.18 Warning sign against laser beams, after [1]. (Image source: https://publicdomainvectors.org)

quantities are the *incident electric field strength* E_{inc} [V m^{-1}], *incident magnetic field strength* H_{inc} [A m^{-1}] and *incident power density* S_{inc} [W m^{-2}]. The latter quantity is used for the determination of reference values for the skin.

ICNIRP has determined the field strengths which, under worst-case conditions, result in the limiting exposures to the whole body or specific tissues as expressed by the basic restrictions. The reference quantities depend on the frequency of the incident radiation and are shown for the occupational exposure of workers in Figs. 2.17 and 2.18.

2.5   Lasers at Accelerators

Lasers emit narrow beams of coherent, monochromatic optical radiation from the far ultraviolet (UV) (180 nm) to the far infrared (IR) (3000μm) range of wavelengths (Fig. 2.18). Lasers have wide-ranging applications in particle accelerator centres, both in the accelerator and in the adjacent laboratories and workshops.

2.5.1   Application of Lasers at Accelerators

The intense light emitted by Lasers has specific applications at accelerators, for example:

- Photo-cathodes illuminated by pulsed laser beams provide electron beams with a specified time structure, matched to the accelerating structures for linear accelerators [34].
- Gamma factories: inverse Compton scattering of laser-generated photons on relativistic electron beams [44] or ion beams [41] provide photons with extremely high energies.
- Laser wakefield acceleration (LWFA). An intense, short Laser pulse traversing a plasma cell-displaces the free electrons, creating an electrical wakefield (Fig. 2.19).

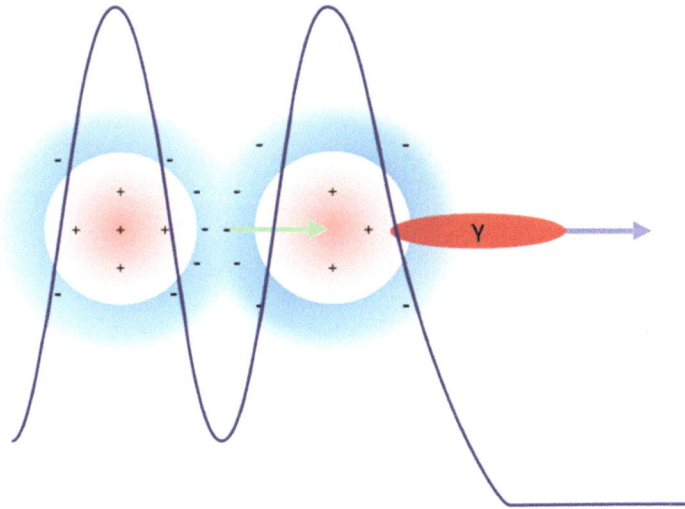

Fig. 2.19 Principle of Laser Wakefield Acceleration: a short intense Laser pulse (Y) traverses a plasma. Electrons are displaced by radiation pressure; the Laser pulse creates a periodic wake field of positively charged ions. The blue line indicates the local positive charge density in the plasma. The electric field (not the ions!) is comoving with the Laser pulse and it accelerates the electrons in its extremely high field gradient. (From [33])

Field gradients surpassing everything that is possible with RF cavities can be generated and electrons have been accelerated to 7.8 GeV over short distances [37]. The lasers used for LWFA have instantaneous powers in the Terawatt range.

- Laser ion source. The sharply defined wavelength of a Laser permits isotope-selective ionisation of exotic atoms in radioisotope facilities, for example ISOLDE at CERN [36]. Several excitation steps are necessary to bring an electron to a loosely bound Rydberg state and finally to remove it from the atom. Tuneable lasers based on toxic dyes are mounted with frequency-doubling crystals and other optical elements on an optical bench (Fig. 2.20).

Unspecific, industrial applications of Lasers in particle accelerator centres are

- Metrology, for example distance measurement, surveying, aligning of accelerator components, laser velocimetry, laser vibrometers.
- Material processing, such as cutting, welding, drilling, photolithography, additive manufacturing ("3D printing").
- Communication, with laser diodes and optical fibres

Fig. 2.20 Optical bench in the ISOLDE RILIS laser laboratory. Operators must protect themselves from direct and reflected laser beams with protection glasses. (Copyright CERN, reused with permission)

2.5.2 Hazardous Effects of Lasers

The collimation of the laser light in narrow beams implies a high power density, expressed in the quantity *irradiance E* with the unit of surface power density [W m^{-2}]. The time integral of irradiance is *radiant exposure J*, an energy surface density [J m^{-2}]. The principal variable to determine the effects of laser light besides its wavelength is imparted energy. For pulsed lasers, radiant exposure is the relevant quantity because irradiance may assume misleadingly high values during the short pulse duration.

The relation between imparted energy and biological effects is strongly dependent on wavelength because different damage mechanisms are at play. The most critical organ for laser damage is the eye (Fig. 2.21). The eye has a physiological reaction to protect itself against intense light, the optical aversion reflex. It is effective for continuous-wave lasers of not too high irradiance. Powerful pulsed lasers may cause irreversible damage to the eye before the aversion reflex is even entering into action. The skin may also be affected when directly irradiated. The biophysical effects of laser radiation on the eye and skin are described in Table 2.6.

Injury thresholds of tissues after exposure to laser light were originally determined in animal experiments. Laboratory animals were exposed to laser beams with specific wavelength, duration, spot size and radiant exposure. After irradiation, damage to the critical parts of the eye (cornea, lens, or retina) was evaluated in an ophthalmological exam. The threshold value for damage is defined as the dose

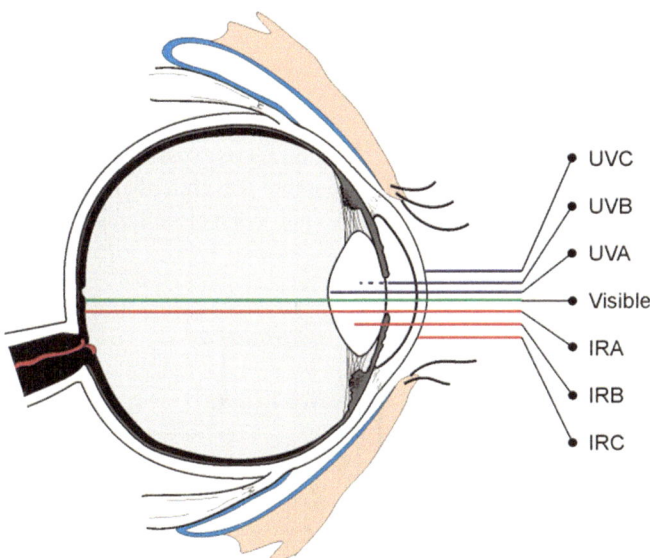

Fig. 2.21 Penetration of different wavelengths through the eye. (With permission, from [35])

Table 2.6 Biophysical effects associated with exposure to laser light beyond the injury thresholds, adapted from [39]

Optical wavelength range	Effect on eye	Effect on skin
UV-C (180 nm–280 nm)	Photokeratitis, Photoconjunctivitis	Erythema (sunburn)
		Skin cancer
UV-B (280 nm–315 nm)	Photokeratitis, Photoconjunctivitis	Erythema, increased pigmentation, skin cancer, Photoageing
	Cataract	
UV-A (315 nm–400 nm)	Photochemical cataract	Increased pigmentation, photosensitive reactions, skin burn
Visible (400 nm–780 nm)	Photoconjunctivitis	
	Cataract	
	Thermal retinal damage	
IR-A (740 nm–1400 nm)	Cataract, retinal burn	Skin burn
IR-B (1.4μm–3.0μm)	Aqueous flare, cataract, corneal burn	
IR-C (3.0μm–1 mm)	Corneal burn	

(radiant exposure) with a 50% probability for minimal injury. The animal data are supported by clinical data from human exposure, either of volunteers, accidental cases, or clinical patients. Today, lasers have many applications in medicine [43] and clinical data are available from ophthalmology and surgery. They are supported by biophysical models for the complex laser-tissue interactions [42]. Such models can be used to extrapolate injury thresholds to exposure conditions which are not experimentally investigated.

Other safety hazards of lasers are:

- Fire hazard from powerful lasers, as their light output can ignite flammable material;
- Chemical hazards from toxic and flammable dyes and solvents employed in the operation of tuneable lasers with adjustable wavelength;
- Electrical hazards from power sources and control systems of lasers.

2.5.3 Protection Against Laser Exposure

2.5.3.1 Maximum Permissible Exposure Limits

In the previous section, injury thresholds for laser exposure were introduced, they are determined from animal experiments, clinical data, and modelling. For operational protection against laser exposure, Maximum Permissible Exposure (MPE) limits are derived by applying reduction factors (typically a factor of 10) to the injury thresholds. Depending on the application for pulsed or continuous-wave

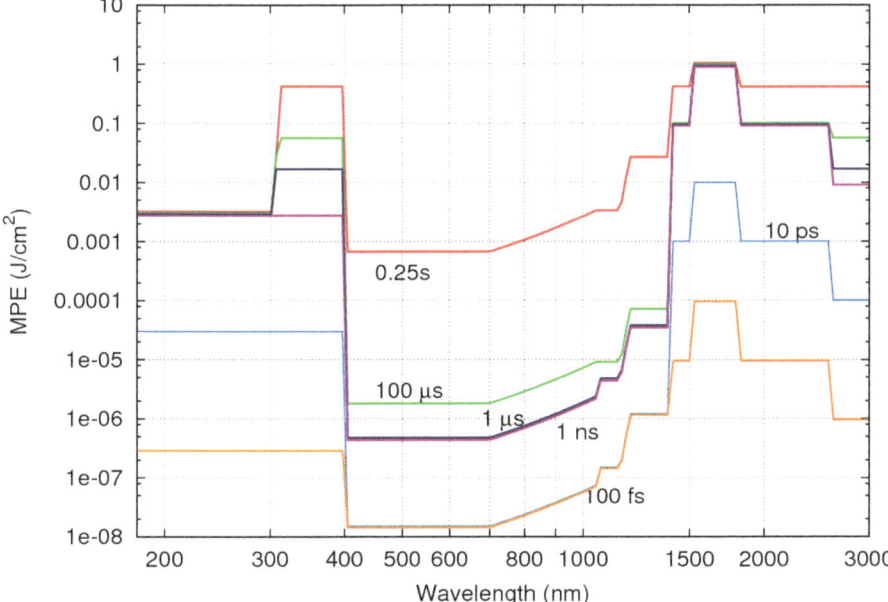

Fig. 2.22 Maximum Permissible Exposure (MPE) limits expressed as radiant exposure [J cm^{-2}] versus wavelength for various exposure times (pulse durations). (From [45])

lasers, MPE are expressed either in the quantity irradiance E (W m^{-2}) or radiant exposure J (J m^{-2}), and they are tabulated as a function of wavelength, exposure duration and spot size in the reference publication of ICNIRP (International Commission for Non-Ionizing Radiation Protection) [38]. Figure 2.22 shows a graphic rendition of the MPE expressed in radiant exposure for different wavelengths and pulse times [45]. The plot shows MPE values varying over several orders of magnitude in short wavelength intervals, reflecting the complex interplay between the laser, and the tissues' physical and physiological reactions.

In a practical situation, a laser specialist must determine the possible exposure scenarios, determine irradiance or radiant exposure for each scenario by measurement or calculation and compare these values with the applicable MPE. This process is complicated but must nevertheless be applied in laser applications in research and development as for example in Fig. 2.20.

2.5.3.2   Laser Classification

For the application of the MPE limits stated by ICNIRP in standard situations, a different approach is chosen. Since one cannot expect that every Laser owner has the expertise to determine exposure values and compare them with the relevant MPE, one classifies Lasers based on their emission capabilities. Classes range from 1 for the least dangerous to 4 for the most dangerous Lasers. Classes may be subdivided to consider special situations. For each class, the maximal potential exposure risk is determined by measurement and calculation, and the Accessible Emission Limit (AEL) is determined for a few conservative exposure scenarios of the eye or the skin. The corresponding protective measures are applied to all lasers belonging to the same class. The classification of a laser is the duty of the manufacturer, who may refer to testing laboratories with the necessary expertise. Classification tests are designed to be rather "worst-case" and restrictive in order to ensure that a "low-class" (e.g. Class 1) laser does not present a hazard to the eye or skin even in reasonably foreseeable worst-case situations. The International Electrotechnical Commission (IEC) publishes the series of standards "Safety of Laser Products ", IEC 60825. The first part of the series [39] introduces the said classification (Table 2.7) and specifies the emission limits for each class, as a function of wavelength and pulse duration. Once classified, a label is affixed visibly on the laser to indicate the class that it belongs to. (Fig. 2.23).

Table 2.7 Laser Safety Classification, adapted from [39] and [40]. Class 1C, for medical and cosmetic applications, has been omitted

Class	Criterion	Potential Health Effect	Mitigation
1	Laser products that are safe during use, including long-term direct intrabeam viewing, even when exposure occurs while using telescopic optics.	Safe under reasonably foreseeable conditions	No measures required ("eye-safe")
	Class 1 also includes high power lasers that are fully enclosed so that no potentially hazardous radiation is accessible during use (embedded laser product).	Intrabeam viewing of Class 1 laser products which emit visible radiant energy may still produce dazzling visual effects, particularly in low ambient light.	Follow manufacturer instruction for safe use
1 M	Laser products that are safe, including long-term direct intrabeam viewing for the naked eye. The MPE can be exceeded following exposure with telescopic optics such as binoculars for a collimated beam.	Safe for naked eye, may be hazardous if the user employs optics.	Laser is localised or enclosed
	The wavelength region for Class 1M lasers is restricted to the spectral region between 302,5 nm and 4000 nm.	Intrabeam viewing of Class 1M laser products which emit visible radiant energy may still produce dazzling visual effects, particularly in low ambient light.	User safety training recommended. Prevent use of magnifying, focusing or collimating optics.

(continued)

Table 2.7 (continued)

Class	Criterion	Potential Health Effect	Mitigation
2	Laser products that emit visible radiation in the wavelength range from 400 nm to 700 nm that are safe for momentary exposures but can be hazardous for deliberate staring into the beam.	Safe for short exposures; eye protection is afforded by aversion response.	Follow manufacturer instruction for safe use.
		Dazzle, flash-blindness, and afterimages may be caused by a beam from a Class 2 laser product, particularly under low ambient light conditions. This may have indirect general safety implications resulting from temporary disturbance of vision or from startle reactions.	Do not to stare into the beam.
2 M	Laser products that emit visible laser beams and are safe for short time exposure only for the naked (unaided) eye. The MPE can be exceeded following exposure with telescopic optics such as binoculars for a collimated beam	Safe for naked eye for short exposures, may be hazardous if the user employs optics.	Laser is localised or enclosed.
			User training recommended
		Dazzle, flash-blindness, and afterimages may be caused by a beam from a Class 2M laser product.	Do not to stare into the beam.
			Prevent use of magnifying, focusing or collimating optics.
3R	Laser products that emit radiation that can exceed the MPE under direct intrabeam viewing.	Risk of injury is relatively low but may be dangerous for improper use by untrained persons.	Laser is enclosed.
			User training required.
			Subject to the findings of the risk assessment, personal protective equipment may be necessary
		The risk of injury in most cases is relatively low. The risk of injury increases with exposure duration.	Prevent direct eye exposure.
3B	Laser products that are normally hazardous when intrabeam ocular exposure occurs including accidental short time exposure.	Direct viewing is hazardous.	Laser is enclosed and interlock protected.
		Eye injury during intrabeam exposure	Key control for access to laser.
			Personal protective equipment required.
	Viewing diffuse reflections is normally safe.	Lasers which approach the AEL for Class 3B may produce minor skin injuries or even pose a risk of igniting flammable materials.	Prevent eye and skin exposure to the beam. Guard against unintentional reflections
			Appoint competent person as Laser Safety Officer (LSO)
4	Laser products for which intrabeam viewing and skin exposure is hazardous and for which the viewing of diffuse reflections may be hazardous.	Eye and skin injury during intrabeam exposure and viewing of diffuse reflections.	Laser is enclosed and interlock protected.
			Key control for access to laser.
		Fire hazard	Personal protective equipment required.
			Prevent eye and skin exposure from direct and diffuse reflection of the beam.
			Appoint LSO

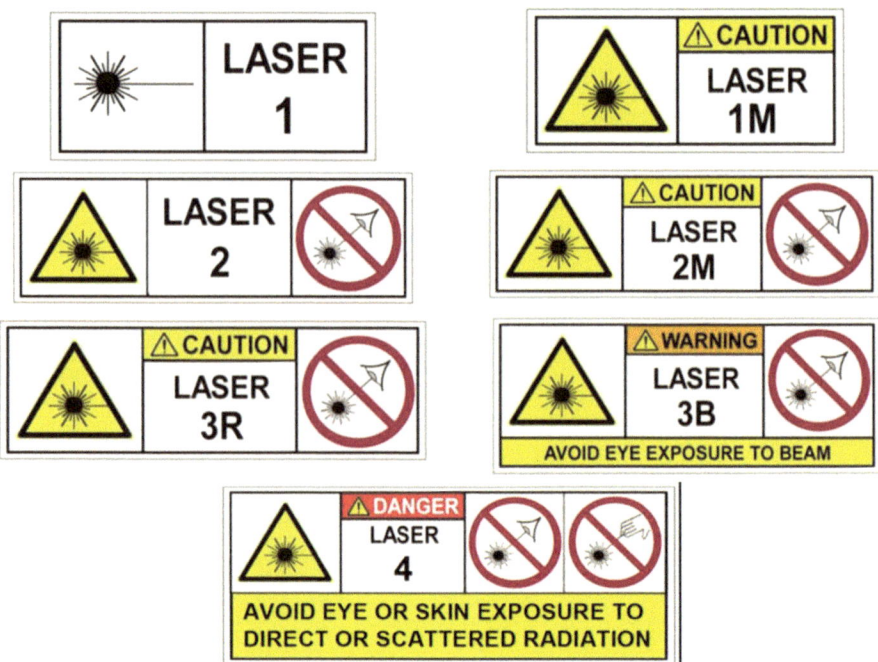

Fig. 2.23 Warning labels to be affixed on lasers, according to class. The hazard mitigation measures are expressed in unmistakable symbols. (From [39]. (Copyright © 2014 IEC Geneva, Switzerland. www.iec.ch)

2.5.3.3 Practical Laser Safety

Following the Hierarchy of Controls (Sect. 5.1.5), the most effective prevention measure against accidents with lasers is isolation so that access to the light path is impossible during operation. Class 1 also includes lasers with higher irradiance or radiant exposure than the AEL, but with a fully encapsulated light path and built in a way that they can only operate if the protection is intact.

If the light path of a laser from a class higher than 2 M cannot be fully enclosed, it must be installed in a laser room in which the entrance door is interlocked with the power source of the laser or an optical shutter, so that the laser beam stops once the door is opened. For classes 3B and 4, a key control for the laser room or for laser operation must be installed, permitting access to and operation of the laser only by trained personnel.

In laser set-ups where the direct beam runs in the open for purposes of adjustment, personal protective equipment (PPE) is the last resort. Hands and forearms can be protected from UV radiation by leather gloves and long sleeves, and the eyesight with goggles, attenuating the characteristic wavelength of the laser.

2.6   Beam-Intercepting Devices

As the name of this section indicates, it treats devices which interact directly with the particle beam by being placed in its path. Beam intercepting devices must be able to handle the energy transmitted to them, mostly in form of heat and ionising radiation. The principal occupational hazard of these devices is activation by interaction with particles (Sect. 3.1) and the resulting exposure of personnel to ionising radiation (Sect. 3.5.2), occurring during periods of maintenance and repair.

2.6.1   Collimators

In simple accelerator theory, an accelerated particle orbits in an accelerator or storage ring with no other interactions than with the electromagnetic guiding and accelerating fields. In reality, dissipative forces from collisions with rest-gas atoms and interactions with other particles make the particles deviate from their ideal orbit. Other sources of orbit errors are the pulsed electromagnetic fields in kickers and septa used to steer the beam from one accelerator into another one. Particles which do not experience the full strength of the "kick" deviate from the ideal orbit. The result of these dissipative and deviating interactions is a growth of the beam-size and an increased probability that beam particles collide with the accelerator components. These collisions are the source of the following adverse effects:

- In accelerators with normal conducting magnets the collisions lead to secondary particle cascades with numerous nuclear interactions and lead to activation of the accelerator components (Sect. 3.3.2).
- In accelerators with superconducting magnets the collisions of charged particles with the magnets lead to energy deposition outside the thermodynamic stability of the superconductor and lead to a quench (Sect. 2.2.2). The operational stability and availability of the accelerator make the avoidance of such beam loss mandatory.

The answer to these two points is to concentrate beam loss in a few locations of the accelerator, where it does not damage equipment or quench magnets, and where secondary particle cascade and the resulting activated material can be shielded [48]. Collimators, consisting of massive absorber blocks, present artificial restrictions in the accelerator beam line to absorb particles steering off the ideal orbit. Very high energy particles are scattered in a large solid angle by the collimator material, thus diluting their damaging effects.

- Radioactivity after use. The dose rate levels around a collimator in a storage ring are in general so high – a few mSv/h in working distance – that the equipment cannot be maintained shortly after the stop of the accelerator beam.

- Radioactive contamination. A collimator operates under high or ultrahigh vacuum conditions. During collimator maintenance the vacuum tank is opened and workers may be exposed to dust and debris from collimator materials which have become brittle by constant particle beam impact. Cooling water or hydraulic fluids in a collimator may be slightly activated, although by experience, this is rarely of concern.
- In a rigorous interpretation of the legislation, the actuators or motors make the collimator a "machine" which falls under the corresponding directives (see 4.2.1). However, equipment for the use in fundamental research may be exempted from the prescriptions of the directive and the local prescriptions should be clarified with the licensing authorities (Fig. 2.24).

One strategy to mitigate high personal radiation dose from activated collimators is to keep a damaged collimator in an intermediate storage and wait until the dose rate is acceptable before attempting a repair. The acceptable level depends on the planned duration of the work (Sect. 3.5). Alternatively, one may decide beforehand that collimators are expendable and not even try to repair them. The design of a collimator contributes also to the optimization of radiation protection: a plug-and-play assembly allowing rapid exchange of components saves time and dose.

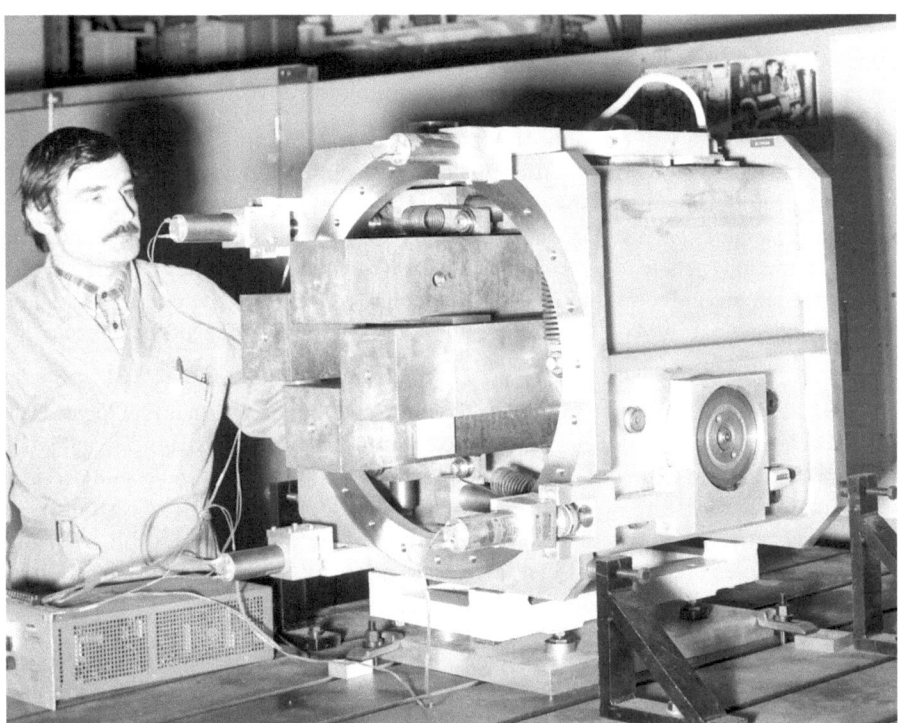

Fig. 2.24 Collimator for extracted beamlines from the SPS at CERN. While not the latest design, this photo shows the metal collimator blocks which, in this model, can be moved to restrict the beam in two directions simultaneously. (Copyright CERN, reused with permission)

2.6.2 Targets

One can distinguish between different types of targets at an accelerator: production targets, experimental targets, and stripping foils. They have in common that they are placed in the path of an accelerated particle beam to provoke beam-matter collisions.

2.6.2.1 Production Targets

Production targets serve to produce unstable particles in collisions with the primary beam. Examples are spallation targets from heavy metals to produce neutrons, or light metal targets to produce antiprotons. Production targets are usually massive objects to optimise the yield of the desired particles.

The yield of a production target is proportional to the number of reactions within the target, which is in turn the product of the beam intensity, the length of the primary particle's path within the target and the production cross section of interest. Besides the reaction of interest, the primary charged particle will have numerous other interactions within the target material, mediated by the electro-magnetic and the strong fundamental interactions. These processes are described with more detail in section 3.2. As a bottom line, most of the energy deposited during the passage of charged particles through matter serves to heat the target material [47], ionisation and energy leaving the target in form of radiation and new particles are often negligible. In spallation sources [5, 8], the neutron production targets cope with several MW of absorbed beam power. An aggravating factor is the pulsed time structure of the beam which leads to even higher instantaneous values of power during the beam impact. This places stringent requirements on the mechanical and thermal resistance of the targets.

2.6.2.2 Experimental Targets

Experimental targets are placed in the path of a particle beam, the properties of the reaction products of collisions (particle type, momentum, spin) are observed with detectors and provide insight to fundamental interaction mechanisms. As a rule, experimental targets are thin to let the reaction products escape without alteration of their properties in secondary collisions.

Experimental targets may consist of radioactive material: to produce superheavy isotopes one bombards targets enriched in unstable transuranium isotopes with beams of the heaviest possible stable or long-lived elements, lead, gold, or the Uranium-isotope ^{238}U. The use of these radioactive targets is subject to authorisation procedures in national regulations. Their handling requires the application of strict administrative procedures. Most of the transuranium isotopes employed in production targets for superheavy isotopes are alpha emitters with a high radiotoxicity. They must be handled with utmost care to prevent internal exposure of personnel.

A special type of experimental target is the instrumented beam dump. It is a thick target, surrounded and followed by detectors to investigate the debris from collisions in the search of long-lived exotic particles.

2.6.2.3 Stripping Foils

Stripping foils are very thin foils (in the range of micrometres) placed in the path of a negatively charged beam to remove orbital electrons. They are used to strip heavy ions from all electrons, or to remove two electrons from the H- ion in an efficient injection process in a synchrotron.

2.6.3 Beam Dumps

Particle beams carry high energy and can damage accelerator equipment by different interaction mechanisms (Sect. 3.2). This energy must be deposited safely once the beam has interacted in an experiment or has become too diluted to provide meaningful collision rates with the opposing beam in a collider. The beam is directed onto a massive absorber where most of its energy is absorbed or dissipated. The high beam energy poses challenges to cooling and mechanical stability of the dumps.

One distinguishes between *external* and *internal* beam dumps. External beam dumps are constructed at the end of a transfer beam line. When the signal for dumping the beam is given, kicker magnets deflect it out of its normal orbit into the transfer beam line. The extraction kicker and transfer beam line are optimised for the transport of particles within a narrow energy band. The LHC is equipped with two external dumps and transfer lines optimised for the terminal energy of LHC, 7 TeV.

If particles with many different energies must be absorbed in a synchrotron (for example, particles shortly after injection and at different terminal energies), an internal dump is the best choice. Internal dumps resemble collimators [46]. Particles can pass through an internal dump by a large beam space without interacting or even being deviated from their orbit (Fig. 2.25). Upstream of the dump, a kicker is mounted which can deflect the particles into the absorber block, where part of their

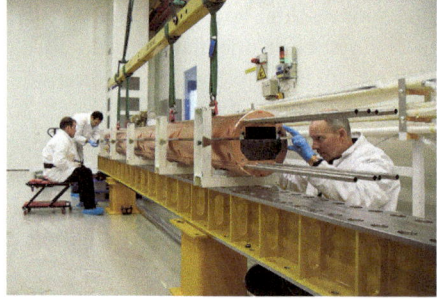

Fig. 2.25 Internal beam-dump of the SPS at CERN. Left: Assembly of the dump core. In the copper cylinder (for mechanical stability and cooling) one distinguishes the absorber block on the bottom and the free space for the particle beam above it. Right: assembly of the dump's radiation shielding in a surface building for test purposes. (Copyright CERN, reused with permission)

energy is absorbed, and part dispersed as a secondary particle cascade. The combination of a kicker followed shortly by an internal dump has a large acceptance for particle energies. An internal dump is mounted in the SPS synchrotron at CERN, where protons are accelerated from 20 GeV to 450 GeV. The internal dump allows beam aborts at any energy.

The hazards of beam dumps are similar to those of collimators. Due to the high number of absorbed particles, the radioactivity of a dump is so high that personnel can only work for a limited time in their vicinity. Internal dumps are surrounded by massive radiation shielding to protect personnel who must pass the dump location or work in its vicinity, external dumps are located in blind tunnels where no other access is required. Because the repair of a dump is possible only after storage for radioactive decay for a few years, one must have spares in store to replace the failed dump.

References

1. The European Council, *Council Directive 92/58/EEC on the minimum requirements for the provision of safety and/or health signs at work* (EEC Brussels, 1992). https://eur-lex.europa.eu/legal-content/EN/TXT/?uri=CELEX:01992L0058-20190726

Accelerators

2. European Organization for Nuclear Research, *LEP design report. Vol II: The LEP main ring*, CERN-LEP/84-01 (CERN, Geneva 1984). http://cds.cern.ch/record/102083
3. O. Brüning, P. Collier, P. LeBrun, S. Myers, R. Ostojic, J. Poole, P. Proudlock (eds.), *LHC Design Report Vol.1 The LHC Main Ring, CERN-2004-003* (CERN, Geneva, 2004)
4. Future Circular Collider Study, https://fcc.web.cern.ch/Pages/default.aspx
5. R. Garoby et al., The European spallation source design. Phys. Scr. **93**, 014001 (2018), at https://europeanspallationsource.se/publications
6. Paul-Scherrer Institute, Laboratory for Particle Physics (LTP), LTP facilities https://www.psi.ch/en/ltp/facilities
7. International Linear Collider, https://ilchome.web.cern.ch/ilc/facts-and-figures
8. Spallation Neutron Source Proton Power Upgrade, https://neutrons.ornl.gov/ppu

Magnets

9. F. Bordry et al., Accelerator Engineering and Technology: Accelerator Technology, in *Particle Physics Reference Library*, ed. by S. Myers, H. Schopper, (Springer, Cham, 2020)
10. J.F. Bouteille, in *Proceedings of the CAS–CERN Accelerator School: Power Converters*, Baden, Switzerland, 7–14 May 2014, ed. R. Bailey, CERN-2015-003 (CERN, Geneva, 2015), https://doi.org/10.5170/CERN-2015-003.171
11. CERN, *The 300 GeV Programme, CERN/1050* (CERN, Geneva, 1972), http://cds.cern.ch/record/24016
12. International Commission on Non-Ionizing Radiation Protection, Statutes of ICNIRP Approved at the Commission Meeting, 13–14 October 2008, in Rio de Janeiro, Brazil, https://www.icnirp.org/cms/upload/doc/statutes.pdf

13. International Commission on Non-Ionizing Radiation Protection, ICNIRP Guidelines on Limits of Exposure to Static Magnetic Fields. Health Phys. **96**(4), 504–514 (2009); https://www.icnirp.org/cms/upload/publications/ICNIRPstatgdl.pdf
14. S. Russenschuck, *Field Computation for Accelerator Magnets* (Wiley-VCH, Weinheim, 2010)
15. S. Russenschuck, *Private Communication* (CERN, Geneva, 2018)
16. Th. Zickler, in *Proceedings of the CAS–CERN Accelerator School: Magnets,* Bruges, Belgium, 16–25 June 2009, ed. by D. Brandt, CERN-2010-004 (CERN, Geneva, 2010), pp. 65–102, https://doi.org/10.5170/CERN-2010-004.65
17. A.V. Zlobin, D. Schoerling, Superconducting Magnets for Accelerators, in *Nb3Sn Accelerator Magnets. Particle Acceleration and Detection*, ed. by D. Schoerling, A. Zlobin, (Springer, Cham, 2019)

Cryogenics

18. H. DeVoe, Thermodynamics and Chemistry 2nd ed., University of Maryland, http://www2.chem.umd.edu/thermobook/ (2020)
19. L. Dufay-Chanat et al., Final report on the Controlled Cold Helium Spill Test in the LHC tunnel at CERN, IOP Conf. Ser. Mater. Sci. Eng. **101** (2015) 012123, https://iopscience.iop.org/article/10.1088/1757-899X/101/1/012123
20. F.J. Edeskuty, W.F. Stewart, *Safety in the Handling of Cryogenic Fluids* (Springer Science and Business Media, New York, 1996)
21. European Industrial Gases Association, *Campaign against asphyxiation*, Safety Newsletter Special Edition SAG NL N° 77/03/E Brussels, (2003) https://www.eiga.eu/publications/safety-newsletters/nl-7703e-campaign-against-asphyxiation/
22. International Organisation for Standardization, ISO 21013 Cryogenic vessels — Pressure-relief accessories for cryogenic service (4 parts, ISO, Geneva, 2007–2016)
23. C.B. Keil et al., *Mathematical Models for Estimating Occupational Exposure to Chemicals*, 2nd edn. (American Industrial Hygiene Association, Fairfax, 2009)
24. Ph. Lebrun et al., Cryogenics for Particle Accelerators and Detectors, CERN LHC/2002-11 (CERN, Geneva 2011) http://cds.cern.ch/record/592467
25. Z.M. Malecha et al., Experimental and numerical investigation of the emergency helium release into the LHC tunnel. Cryogenics **80**, 17–32 (2016). https://doi.org/10.1016/j.cryogenics.2016.09.005
26. K.D. Timmerhaus, T.M. Flynn, *Cryogenic Process Engineering* (Springer Science and Business Media, New York, 1989)
27. S.W. Van Sciver, *Helium Cryogenics*, 2nd edn. (Springer, New York/Dordrecht/Heidelberg/London, 2012)

RF Technologies

28. Proceedings of the CAS-CERN Accelerator School: Basic Course on General Accelerator Physics, Loutraki, Greece, 2–13 October 2000, edited by F. Ruggiero and J. Thomashausen, CERN-2005-004. https://doi.org/10.5170/CERN-2005-004
29. Proceedings of the CAS–CERN Accelerator School: RF for accelerators, Ebeltoft, Denmark, 8–17 June 2010, edited by R. Bailey, CERN-2011-007. https://doi.org/10.5170/CERN-2011-007
30. International Commission on Non-Ionizing Radiation Protection, ICNIRP Guidelines for limiting Exposure to Electromagnetic Fields (100 kHz to 300 GHz); Health Physics 118 (2020) and https://www.icnirp.org/cms/upload/publications/ICNIRPrfgdl2020.pdf

31. J.D. Jackson, *Classical Electrodynamics*, 3rd edn. (Wiley, New York, 1998)
32. D. Zhou et al., Radiation shielding study for the vertical test system for superconducting Rf cavities. Rad. Prot. Dosim **185**, 124–130 (2019)

Lasers

33. A. Cakir, A.O. Guzel, A Brief Review of Plasma Wakefield Acceleration, arXiv:1908.07207v3 [physics.acc-ph] (2019)
34. E. Chevalley et al., Photo-cathodes for the CERN CLIC Test Facility, CERN-PS-98-036-LP; CLIC-Note-373, http://cds.cern.ch/record/364521
35. European Union, Non-Binding Guide to Good Practice for implementing Directive 2006/25/EC 'Artificial Optical Radiation', (Brussels: 2011), https://op.europa.eu/en/publication-detail/-/publication/556b55ab-5d1a-4119-8c5a-5be4fd845b68
36. V. Fedosseev et al., Ion beam production and study of radioactive isotopes with the laser ion source at ISOLDE, J. Phys. G: Nucl. Part. Phys. **44**, 084006 (2017), http://cds.cern.ch/record/2291843
37. A. Gonsalves et al., Petawatt laser guiding and electron beam acceleration to 8 GeV in a laser-heated capillary discharge waveguide. Phys. Rev. Lett. **122**, 084801 (2019)
38. International Commission on Non-Ionizing Radiation Protection, ICNIRP Guidelines on limits of Exposure to Laser Radiation of Wavelengths between 180 nm and 1000 µm. Health Phys. **105**(3), 271–295 (2013); https://www.icnirp.org/cms/upload/publications/ICNIRPLaser180gdl_2013.pdf
39. International Electrotechnical Commission, *Safety of Laser Products –Part 1: Equipment classification and requirements, IEC 68025-1:2014* (IEC, Geneva, 2014)
40. DIRECTIVE 2006/25/EC of the European Parliament and of the Council on the minimum health and safety requirements regarding the exposure of workers to risks arising from physical agents (artificial optical radiation). http://data.europa.eu/eli/dir/2006/25/2019-07-26
41. W. Krasny, The Gamma Factory Proposal for CERN Geneva (2015), https://arxiv.org/abs/1511.07794
42. K. Kulikov, *Laser Interaction with Biological Material – Mathematical Modelling* (Springer, Cham/Heidelberg/New York/Dordrecht/London, 2014)
43. M.H. Niemz, *Laser-Tissue Interactions* (Springer Nature Switzerland AG, Cham, 2019)
44. H. R. Weller et al., Gamma Beam Delivery and Diagnostics, Romanian Reports in Physics, Vol. 68, Supplement, S447–S481 (Bucharest 2016), http://www.rrp.infim.ro/2016_68_S/S447.pdf
45. Figure in article "Laser Safety", https://en.wikipedia.org/wiki/Laser_safety

Beam-Intercepting Devices

46. A. Pilan Zanoni, Characterization and core renovation of beam stoppers for personnel safety. J. Instrum. **14**, T01011 (2019). https://doi.org/10.1088/1748-0221/14/01/T01011
47. M.A. Plum, *Proceedings of the Joint International Accelerator School: Beam Loss and Accelerator Protection, Newport Beach, United States, 5–14 November 2014, edited by R. Schmidt, CERN-2016-002* (CERN, Geneva, 2016), pp. 229–251. https://doi.org/10.5170/CERN-2016-002.229
48. S. Redaelli, *Proceedings of the Joint International Accelerator School: Beam Loss and Accelerator Protection, Newport Beach, United States, 5–14 November 2014, edited by R. Schmidt, CERN-2016-002* (CERN, Geneva, 2016), pp. 403–437. https://doi.org/10.5170/CERN-2016-002.403

Chapter 3
Beam Hazards and Ionising Radiation

Abstract This chapter treats hazards originating from particle beams. The interaction of charged particle beams with matter is described. Beam loss can cause material damage in structural and electronic components. Ionising radiation is introduced by a description of the different types of radiation. Then, the sources of ionising radiation at accelerators are defined: beam loss is the origin of prompt ionising radiation. Material activated by the passage of particle cascades is a long-lived source of ionising radiation. The chapter is closed with a description of radiation dosimetry and radiation protection at accelerators.

3.1 Beam Loss in Particle Accelerators

Ideally, a particle accelerator works with 100% efficiency, transporting all charged particles from a source to a target while giving them the required total energy. In reality, collisions between beam particles and matter are mostly unwanted, but unavoidable. With increasing energy and intensity of the particle beams delivered, the effects of particle losses ("beam loss") are becoming more pronounced. As an example, the CERN Proton Synchrotron (PS) accelerated in the early 1960s proton beams with an intensity of 10^{12} particles per pulse. In the years around 2000, the beam loss at the extraction septum towards the Super Proton Synchrotron (SPS) exceeded this value by a factor of 2–3 [1].

Sources of unwanted beam-matter interactions, leading as a result to deviations from the ideal orbit, are [9]:

- Imperfect beam line vacuum making beam particles collide with rest gas molecules. This interaction deviates the beam particle from its longitudinal course and may break-up the rest gas particle.
- Electromagnetic fields induced by high-intensity beams increase the transverse momentum of beam particles, and lead to their eventual loss by collision with the beam tube.
- Beam steering elements, such as kickers and septa. Some particles inevitably impinge on the *blade*, a thin foil in a septum defining the electric field which brings the particles on a new course. Another loss mechanism is misfiring of a

© The Author(s) 2021
T. Otto, *Safety for Particle Accelerators*, Particle Acceleration and Detection,
https://doi.org/10.1007/978-3-030-57031-6_3

kicker's electromagnetic circuits, the "kicked" particles will go astray and collide with the beam line at positions which can be predicted from beam dynamics.

• Close encounters of colliding beams in the centre of a physics detector, inducing transversal momentum, leading to small deviations from the "perfect" particle orbit. In circular colliders, these minute deviations may lead to the loss of the particles many turns after the interaction.

In Sect. 2.6, controlled interactions with the beam were introduced:

• Collimators, skimming particles with transversal excursions at predefined locations.
• Beam-dumps, absorbing the particle beam after the target or after it has become too diluted to be useful for further exploitation in a collider.
• Finally, fixed targets in physics detectors for extracted beams are obviously sources of beam-matter interactions.

The first two beam loss mechanisms are distributed over the whole length of the accelerator while all others are localised.

A rare hazard is the direct exposure of personnel to the particle beam. In modern accelerator facilities the risk of this happening is reduced to near zero by accelerator safety systems and access control systems, described in Sect. 5.3. These systems make it near-impossible for a person to enter the accelerator area when particle beams are produced.

3.2   Beam-Matter Interaction

Particles lost from the beam collide with material from the accelerator. Given sufficient energy, they traverse the beam line vacuum tube, the magnets and other equipment. Their path ends in the shielding walls or in the ground or air surrounding the accelerator. This section describes the passage of particles through matter. The multiple interactions between beam and matter generate the radiation fields at an accelerator during operation. They are also responsible for activation of the material. Here, a qualitative overview of this complex subject is given, without going in the physical details and their mathematical description. For more detailed descriptions, one can refer to [3, 6, 7].

3.2.1   Electrons and Positrons

The following interaction mechanisms contribute to the energy loss of electrons and positrons in matter, and to the emission of ionising radiation.

- *Ionisation.* An inelastic collision of the incoming electron with a bound electron mobilises the latter as a so-called δ-electron. The δ-electrons can in turn ionise further atoms on their path.
- *Bremsstrahlung.* The electron emits X-ray photons during deceleration or deviation.
- *Pair production* (PP). A photon with an energy $E > 2\ m_0c^2 = 1.022$ MeV can convert to an electron-positron pair in the electric field of an atom.
- Positrons at rest annihilate with an electron into two or three photons.

Low-energy electrons interact predominantly by ionisation. At the *critical energy* E_c, the probability for ionization and Bremsstrahlung are equal, at higher energy, Bremsstrahlung prevails. The value of critical energy is approximately

$E_c(\text{MeV}) = \dfrac{800}{Z+1.2}$, where Z is the atomic number of the target material [10].

Radiation length X_o is the depth in matter in which Bremsstrahlung has reduced the energy of a high-energy electron ($E \gg E_c$) to $1/e$ of the original value. The energy dependence of the different interaction mechanisms is illustrated for lead in Fig. 3.1.

In thick absorbers, electrons or photons with an energy well above the critical energy E_c initiate an electromagnetic (EM) shower or cascade. In an EM cascade, the interaction mechanisms of compton scattering, pair production and bremsstrahlung alternate and lead to several generations of electrons, electron – positron pairs and high-energy photons. The kinetic energy of the particles decreases

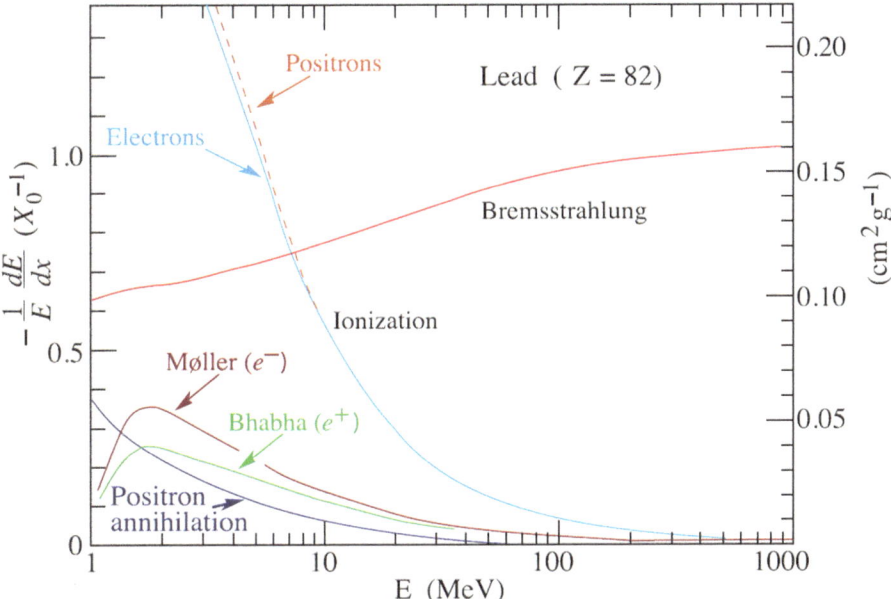

Fig. 3.1 Fractional energy loss per radiation length in lead for each of the interaction mechanisms as a function of electron or positron energy. The critical energy E_c is situated where the curves for Ionization and for Bremsstrahlung cross. (From [8])

in each generation. The cascade stops propagating once the average energy of the electrons has fallen below the critical energy E_c. From this moment on, the probability that the electrons ionise atoms is higher than that of Bremsstrahlung emission. The cascade loses one of its driving particles because the electrons now dissipate their energy preferably by ionisation and come quickly to rest. A schematic illustration of an EM cascade is given in Fig. 3.2.

3.2.2   Protons and Charged Heavy Particles

Protons and charged ions traversing matter are subject to the following interaction mechanisms:

- *Electromagnetic interaction with target electrons.* Charged particles interact with the atomic electrons and nuclei of the target material. Electronic energy loss is the most probable mechanism, because of the mass difference between the projectiles only small amounts of energy in the range of eV can be transferred. This energy leads to excitation of the target atoms and electrons in rotational and vibrational states and finally to heating of the material. Target atoms may also be ionised by the ejection of a δ-electron.
- *Electromagnetic interaction with target nuclei.* These interactions are less frequent than with electrons, but more energy is transmitted. The results of these collisions are a change of direction of the projectile and possibly a displacement of the target atom.

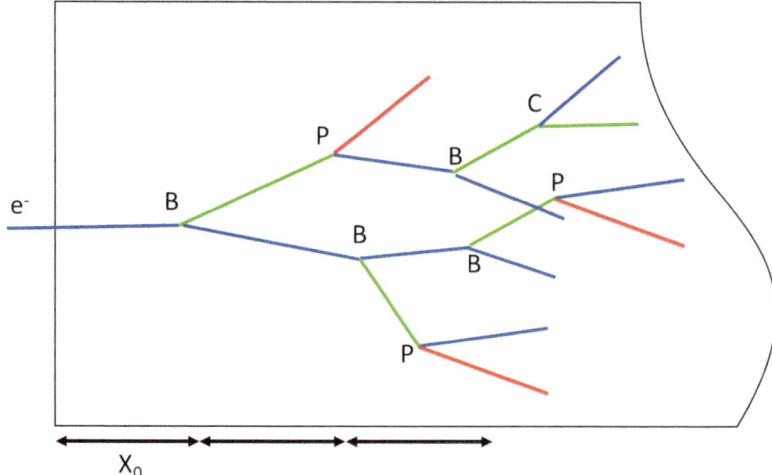

Fig. 3.2 Schematic development of an electromagnetic cascade from a high-energy electron ($E >> E_c$). Electron tracks are blue, positrons red and photons green. Interactions: *B* Bremsstrahlung emission, *P* pair production, *C* Compton scattering. (Image by the author, after [10])

- *Nuclear interactions with target nuclei.* Once a positively charged projectile overcomes the Coulomb-barrier of the nucleus, (typically with a kinetic energy of a few MeV) it can penetrate the nucleus and trigger nuclear interactions leading to a rearrangement of its constituents. Excess energy is carried away by protons, neutrons, or small nuclear fragments (nuclei of ^{2}H, ^{3}H, ^{3}He, ^{4}He, …) leaving the nucleus. This is called the *spallation process.* The kinetic energy of the spallation nuclei and fragments depends on the kinetic energy of the projectile and the released binding energy of the nucleus. Energetic spallation nuclides trigger further electromagnetic and nuclear reactions and a hadronic cascade may be the result.
- *Hadronic interactions in target nuclei.* When the energy of the projectile is high enough, new particles can be generated in nuclei. Pions are produced at energies from 290 MeV on, other, heavier hadrons require correspondingly more energy. The generated particles are unstable, and they decay into stable particles. Charged pions decay to muons and neutrinos and neutral pions to photons. Together with the spallation nuclei and fragments, all these particles form part of the hadronic cascade.

3.2.3   Neutrons

Neutrons are produced at accelerators in two processes:

- *Nuclear and hadronic interactions of charged particles.* As seen above, neutrons are emitted in the spallation process and from hadronic interactions in target nuclei.
- *Photonuclear reactions.* As the name suggests, photons can trigger nuclear reactions, which often result in the emission of a *photoneutron.* The effect is enhanced for heavy target materials with high atomic number Z. In hadronic cascades, and thus in proton or ion accelerators, their rate is negligible against neutron production in the hadronic cascade. In electron accelerators with primary energy above 15 MeV, the production of photoneutrons cannot be neglected. This includes electron linacs for radiotherapy, where heavy metal beam collimators are a source of photoneutrons. They must be considered in the shielding design of such accelerators and they contribute to the unwanted dose in the patient, which is not directed to the tumour.

Free neutrons have a lifetime of about 15 min, they are neutral and can traverse thick shielding walls, and they have a spectrum of effects which spans many orders of magnitude in neutron energy. While charged particles and photons are efficiently attenuated by heavy materials, neutrons are dominating radiation dose and effects outside of particle accelerator shielding (Fig. 3.3).

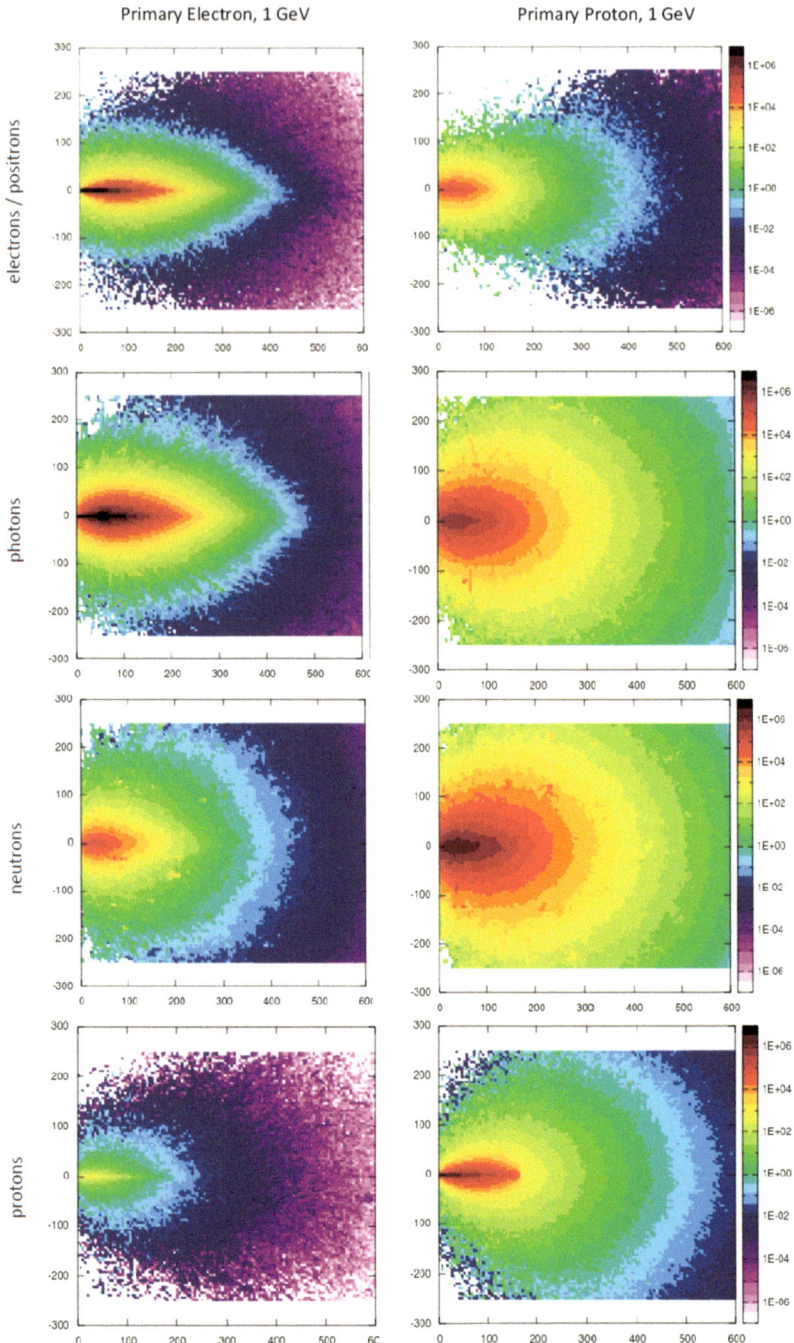

Fig. 3.3 Components of a secondary particle cascade developing in concrete. The primary particles are electrons of 1 GeV, triggering an EM cascade (left column) or protons of 1 GeV, triggering a hadronic cascade (right column). The plot shows the fluence of secondary particles released by 10^9 primary particles hitting the wall from the left. Note that the fluence of secondary particles in a hadronic cascade is several orders of magnitude higher than in an EM cascade. (Image by the author)

3.2.4 Radiation Damage

Radiation damage summarises several effects which are changing the properties of material struck and traversed by radiation or particle beams [2, 5]:

- *Electromagnetic interactions* of projectiles with the target nucleus can lead to atomic displacement. In solid materials atomic displacements lead to crystal defects. A high density of displacements may lead to embrittlement, altering the mechanical resistance of the material. This is of primary concern in high-intensity production targets (Sect. 2.6.2).
- *Nuclear interactions* with target nuclei lead to its nuclear transmutation. The electronic structure of the produced atom does no longer match the crystal lattice and represents another type of lattice defect.
- *Heating* occurs in target materials struck by a particle beam, because the kinetic energy deposited by interactions of the projectile with the target electrons cannot be removed immediately. In extreme cases, local melting may occur, leading eventually to the destruction of the equipment struck by the beam.

These effects are noticeable at places in the particle accelerator where beam loss is prominent: in beam dumps, targets and collimators (Sect. 2.6). Other equipment are the electromagnetic septa used to change the direction of particle beams. These components must be built robust enough to withstand the effects of particle impact. Sometimes, they are built to survive a certain amount of beam before they are preventively exchanged.

Electronic components suffer from three effects mediated by particle beams and hadronic or EM cascades [11]:

- *Single event effects* have their origin in the deposition of a minute amount of charge in an integrated circuit (IC). The charge may flip a logical gate in the IC and provoke an unexpected behaviour. This is a stochastic effect and its probability is proportional to the fluence of secondary particles. Single-event effects are prominent in highly integrated microelectronics, where very small amounts of charge are sufficient to provoke a change.
- *Total Ionisation dose* is the continuous accumulation of absorbed dose in semiconductor materials. Ionisation leads to the creation of electron-hole pairs. In conducting materials, they rapidly recombine, but in the insulating oxides of a semiconductor, holes may persist and, at higher density, permanently change the electronic characteristics of the material.
- *Displacement damage* in silicon, mainly provoked by neutrons. Displaced silicon atoms leave gaps in the crystal lattice, changing its electronic properties.

3.2.5   *Activation of Matter*

Energetic protons cause spallation reactions, having as products nucleons, light nuclear fragments, and a remaining rest nucleus, unstable most of the time. Electron beams can activate matter by photonuclear reactions. The previously non-radioactive material has become radioactive by the exposure to the particle beam.

Estimates for activation are difficult to make, they depend on the energy of the projectile, and the composition of the target material. Rules-of- thumb and approximate formulas are given for electron accelerators by [10] and for proton- accelerators by [34] and [33]. Activity estimates with higher precision are possible with Monte-Carlo radiation transport programs coupled to a nuclear model. Their activation estimates compare with experiment within a factor of two for "good cases" [4] and may be off by an order of magnitude for other nuclei. The prediction of dose rates from activated material by this method is often acceptable because of cancellation effects between nuclei with over- and under-estimated activity.

A frequently repeated rule-of-thumb states that "hands-on" maintenance at a particle accelerator is possible, if the power of beam loss (the product of particle energy and current of lost particles) does not exceed the value of 1 W per metre. This "rule" for which the original source could not be identified, must be applied with care. It originates in the 1960s or 1970s and experience shows that the resulting activation and dose equivalent rate of the material is frequently too high according to modern standards.

3.3   Ionising Radiation

The term "ionising radiation" assembles several physical phenomena that have in common that they can ionise matter. The warning signs against ionising radiation contain a stylised image of a radioactive source (Fig. 3.4).

Fig. 3.4 Warning sign against ionizing radiation and marking of radioactive material transport, after [13], [29]. (Image source: https://publicdomainvectors.org)

3.3.1 Types of Ionising Radiation

Two main families of ionising radiation can be distinguished: directly ionising radiation, mediated by charged particles, and indirectly ionising radiation, by neutral particles and photons.

3.3.1.1 Directly Ionising Radiation

Directly ionising radiation consists of charged particles. They collide with atoms and ionise them by the electromagnetic interaction. One characteristic of directly ionising radiation is that they have a definitive range in matter. Directly ionising radiation may consist of

- alpha (α) radiation, helium nuclei emitted by heavy radioactive nuclei,
- beta (β) radiation is a manifestation of the weak interaction in unstable nuclei, which de-excite under emission of an energetic electron or positron.
- In particle accelerators, other directly ionising particles are produced in collisions between particles and matter, for example muons, pions and protons.

3.3.1.2 Indirectly Ionising Radiation

Indirectly ionising radiation does not interact directly with the atomic electrons. In an intermediate step a charged particle is generated (often a *secondary electron*) which in turn ionises matter. Due to this interaction mechanism, indirectly ionising radiation is exponentially attenuated in matter and has no finite range. One distinguishes between two fundamental families:

- X- and gamma (γ) rays are energetic photons, quanta of the electromagnetic radiation with energy $E > 1$ keV. Photons interact with atoms by the photoelectric effect and the Compton effect, releasing a secondary electron. This secondary particle imparts its energy by electron-atom collisions to matter. Photons with very high energy ($E > 6$ MeV) can interact with the atomic nucleus (photonuclear effect). The excited nucleus will de-excite by emitting either other photons or particles or, for heavy elements, undergo a fission reaction (photofission).
- Neutrons (n) are neutral, nuclear particles emitted by some heavy nuclei, or released in particle – matter interactions. Neutrons interact with atomic nuclei, either scattering elastically (preservation of total kinetic energy, i.e. the internal state of the target nucleus remains unaffected) or inelastically (nuclear reactions take place in the target nucleus). Charged secondary particles are emitted by the excited target nucleus, and a recoil nucleus remains, moving with an energy determined by the reaction kinetics. Recoil protons can take a large fraction of the neutron's energy, because they have nearly the same mass. Certain nuclides absorb thermal neutrons having energies below 0.025 eV. They become activated, with half-lives ranging between fractions of a second and many decades of years.

3.3.2   Sources of Ionising Radiation at Accelerators

Due to the high concentration of energy available in accelerated particles, a large variety of ionising radiation types with wide energy distributions can occur at accelerators. This distinguishes accelerators from facilities in the nuclear power generation cycle, from non-destructive testing and from (most) medical applications.

Two phases can be distinguished with respect to ionising radiation in a particle accelerator:

- During accelerator operation, beam loss, beam-target and beam-beam collisions lead to radiation spectra with an exceptionally wide range of particles and energies, elsewhere encountered only in cosmic radiation. Ionising radiation produced in particle-matter interactions during accelerator operation is also named prompt radiation.
- During maintenance phases, personnel may be exposed to activated material and, exceptionally, to radioactive contamination.

3.3.2.1   Prompt Radiation

At electron accelerators, prompt radiation originates from the electromagnetic cascade, consisting of electrons, positrons, and photons. If the energy of the primary electron is below $E < 10$ MeV, then only these particles contribute to the radiation field. This is the case in medical linacs in diagnostics and therapy in a low-energy setting. Once the threshold of approximately 10 MeV is passed, Bremsstrahlung photons can trigger reactions in nuclei in the so-called photonuclear effect, with production of neutrons. At higher energies, hadrons accompany the EM cascade.

In proton accelerators, the prompt radiation field is described by hadronic cascade, with protons, neutrons, photons, electrons, and positrons as the prevalent components of the radiation field.

An important characteristic of prompt radiation is, that it ceases with the stop of the accelerator. In a properly built accelerator facility, the prompt radiation is correctly shielded by constructing the accelerator underground, or by protecting it with shielding walls made from iron and concrete. It represents in general no risk for workers or the population.

3.3.2.2   Radiation from Activated Material

Ionising radiation emitted from activated material is characteristic for the radionuclides generated, it consists mainly of photons and electrons, with kinetic energies below a few MeV. Different from prompt radiation at accelerators, it does not cease with the stop of the accelerator. Since prompt radiation from an operating particle accelerator is well shielded, exposure to ionising radiation emitted by activated material constitutes the largest contribution to personal dose at an accelerator.

Radiation dosimetry and radiation protection against the exposition to activated material use the same detectors, dosimeters, and procedures as the well-developed field of radiation protection in nuclear or medical facilities.

A particular type of activated material are radiopharmaceuticals, which are produced in small accelerators from stable isotopes by proton bombardment. The produced radionuclides are collected within targets or on foils and they often represent a radioactive contamination hazard. This describes the danger of transferring loose radioactive materials from surfaces to the body and of eventually ingesting or inhaling them. This hazard can be prevented by application of industrial hygiene measures, as they are also employed in workplaces with chemical hazards: protective clothing, Respiratory protection with filters or active air supply. The relative advantage of radioactive contamination over chemical hazards is, that ionising radiation can be detected more easy than chemical contamination.

3.4   Radiation Dosimetry at Accelerators

3.4.1   Dose and Dose Equivalent

The measurement of the detrimental effect of ionising radiation is based on a physical quantity, *absorbed dose D*. It is defined as the ionising energy imparted on a small target of mass δm:

$$D = \frac{\delta \epsilon}{\delta m}$$

In the SI system, the quantity absorbed dose D is measured in the unit J kg^{-1}, which receives the special name Gray (Gy).

One can distinguish two effects of ionising radiation on humans:

- *Tissue reactions* were previously called *deterministic effects* [15] because they appear once a subject has been exposed to an absorbed dose above a threshold. The mildest form is reddening of the skin (erythema), from approximately 0.5 Gy applied to a limited portion of the skin. The thresholds for tissue effects are rather high and they play a secondary role in occupational radiation protection.
- *Stochastic effects* of ionising radiation: these are caused by modifications of the genetic material in cells, either by direct interaction with ionising radiation, or by changes in the cell medium impacting. The probability of developing cancer increases with the radiation dose received, without lower threshold for the effect. The effect is cumulative for chronic exposure.

After World War 2, systematic health studies of survivors of the nuclear bomb explosions over Hiroshima and Nagasaki started, complemented by the follow-up of patients having received radiation treatment. The epidemiological observations are combined with continuously refined biophysical models of exposure to ionising radiation. This work culminated in the recommendations of the International Commission on Radiological Protection (ICRP), with ICRP Publication 103 [14] the most recent in the series.

In the present discussion, tissue effects are neglected, their occurrence at accelerators is extremely rare. In its recommendations, ICRP defines a *protection quantity* for stochastic effects of ionising radiation. It is based on absorbed dose D, but it takes into account the difference in radiation sensitivity of tissues and organs in the body, and the effectiveness of different radiation types to cause cancer. This leads to the quantity *effective dose E*. Effective dose E is not a purely physical quantity, but it includes a measure of the probability to develop cancer after being exposed to ionizing radiation:

$$E = \sum_T w_T \sum_R w_R D_{T,R}$$

In this formula, $D_{T,R}$ is the absorbed dose in averaged over a tissue (organ) T by radiation type R. w_R and w_T are the radiation- and tissue weighting factors, respectively. Effective dose, by definition, is an average of dose over the whole body and cannot be measured directly.

The sum $\sum_R w_R D_{T,R}$ is called *equivalent dose*, it accounts for the radiation effect on a single tissue (organ) T. The physical quantity to express equivalent dose and effective dose is J kg^{-1}. To distinguish it from absorbed dose, it receives the special name Sievert (Sv).

The paradigm of radiation protection is, that the *probability* of radiation detriment to a subject is proportional to the amount of effective dose E received, without a lower threshold. Radiation detriment includes the probability of developing cancer and the ensuing loss of years of life and quality of life. Effective dose limits, limiting the probability of the occurrence of stochastic effects, are determined by comparison with the risk of incapacitating or fatal accidents in other industries. An acceptable level of risk is determined and expressed in effective dose E, which is periodically reviewed by the ICRP. Limits are also set to prevent tissue reactions, with a large safety margin below the threshold dose for reactions (Table 3.1).

To provide measurable quantities, the International Commission on Radiological Units and Measurements (ICRU) has defined so-called *operational quantities*, approximating effective dose. They receive the generic name *dose equivalent* with symbol H and, like effective dose, are quantified in the SI system with the unit J kg^{-1} with the special name Sievert (Sv). The ICRU introduced several operational quantities with specific definitions to cover different exposure situations, for example of the eye lens or the skin. The most commonly used operational quantities are those for the exposure of the full body with $H_p(10)$ (spelled "H-p-10") for the calibration of personal dosimeters and $H^*(10)$ ("H-star-10") for survey instruments.

In contrast to effective dose E, dose equivalent quantities are defined in such a way that they can be realized in standard laboratories and measured with calibrated instruments, within uncertainties.

Table 3.1 Dose limits for ionizing radiation [14]

Type of limit	Occupational	Public
Effective dose	20 mSv per year, averaged over a period of 5 years	1 mSv in a year
Annual equivalent dose in:		
Lens of the eye	20 mSv per year, averaged over a period of 5 years [15]	15 mSv
Skin (averaged over 1 cm^2)	500 mSv	50 mSv
Hands and feet	500 mSv	–

3.4.2   Practical Radiation Dosimetry at Accelerators

The radiation fields around a working particle accelerator are consisting of different types of ionising radiation, extending over broad ranges of energies. This makes dose equivalent measurements at accelerators challenging, and special instruments have been developed over the years to cope with the characteristics of these fields. In shutdown periods, when the accelerator is stopped, the origin of radiation is activated material, containing relatively long-lived radioactive nuclei (with half-lives of days to years), created in beam-matter interactions during the preceding operation periods of the accelerator. These nuclei emit gamma (photon) and beta (electron/positron) radiation. Dose equivalent of these radiations can be measured with standard radiation protection dosimeters and survey instruments, as developed for dosimetry in nuclear and medical radiation facilities.

In a radiation protection program at an accelerator facility, radiation measurement is essential for the following tasks

- measuring dose equivalent at workplaces to make a prospective assessment of working conditions, and to demonstrate the adequacy of the radiation shielding put in place;
- monitoring the dose equivalent to personnel, to demonstrate compliance with legal dose limits;
- monitoring radiation emitted to the environment, to demonstrate compliance with emission and immission limits (Sect. 4.6.1).

Dosimeters are used to measure dose or dose equivalent. In this section, the term *dosimeter* is used likewise for passive and active devices whose purpose is the measurement of dose equivalent. Based on their area of usage, they are may be *personal dosimeters*, *radiation survey instrument* or *radiation monitors*.

One distinguishes between passive and active dosimeters.

- Passive dosimeters are based on a physical effect which is (roughly) proportional to dose equivalent. The magnitude of the physical effect creates a proportional signal once they are read-out after a defined period of exposure. They are used when the immediate display of a result is not essential, for example for personal dosimetry at workplaces with a moderate dose equivalent rate, or for long-term monitoring of the environment.

- Active dosimeters have a detection mechanism which can be directly converted to an electrical signal. Active dosimeters have the advantage to give an immediate display of the dose equivalent, but they are usually larger and more expensive than passive dosimeters. They are used in the form of *electronic personal dosimeters* to monitor workers at workplaces where the dose equivalent rate is so high that they could accumulate a significant fraction of the dose limit in a short time. Active dosimeters are also used for radiation surveys, and as alarm monitors where sudden increases of the radiation intensity are possible and Survey instruments and active radiation monitors usually indicate the instantaneous *dose equivalent rate* $\dot{H} = \dfrac{dH}{dt}$.

Many detectors based on different physical principles exist for the quantification of ionising radiation. Only few of them can be employed as a radiation dosimeter. To make a good dosimeter, the energy-dependent response of the detector to a physical quantity describing the radiation field (for example, particle fluence) must be approximately proportional to the energy dependence of the operational quantity to be measured, for example $H^*(10)$. Hardly any physical radiation detector shows spontaneously the required proportionality of the physical effect to dose equivalent to be used as a radiation dosimeter. It is necessary to modify the energy response of the physical detector to mimic the energy dependence of the dose equivalent quantity to be measured.

This section can only give a very short overview about radiation detection and dosimetry. [20] or [19] give a more detailed treatment of radiation detection, whereas [22] is still a standard reference for radiation dosimetry. [12] gives a modern account of this subject.

3.4.2.1 Photon (Gamma) Dosimeters

The largest fraction of dose equivalent to which workers are exposed in accelerator facilities comes from photons and is accumulated during shutdown periods when the accelerators are maintained. The source of the photons is material activated by interactions with particles during the preceding operational periods. The isotope content of activation products in accelerator facilities is different from those in the nuclear industry by type and concentration, but the same methods of detection and measurement can be employed. Long-lived activation products emit gamma (photon) and beta (electron or positron) radiation.

Photons are indirectly ionising particles, as a first step to detection a secondary electron must be created by the photoelectric effect or by Compton scattering. Most active photon dosimeters rely on one of two detection mechanisms: gas ionisation and creation of free charges in semiconductors. Passive photon dosimeters use thermoluminescence (TL) and optically stimulated luminescence (OSL) as detection principles. Photographic film has been replaced by these technologies in most applications.

Gas-Filled Photon Detectors

Three different types of gas-filled detectors for photons are used in radiation protection: ionisation chambers, proportional counters, and Geiger-Müller counters. The commercially available types have in common a cylindrical geometry with a collection anode in the centre (in proportional counters and GM counters, an anode wire) (Fig. 3.5). Secondary electrons are created in the detector wall and registered in the filling gas by ionisation. The average energy to create an electron-ion pair in gases, the ionisation energy, lies between 20 eV and 40 eV, for air it is approximately 32 eV. The charges are separated by an electrical field between the anode and the chamber wall, and electrons drift to the central anode. The chamber constitutes a cylindrical capacitor for which the radial component peaks at the anode wire.

The three types of gaseous detectors can be distinguished by applied voltage V_0 between chamber wall (cathode) and central anode:

- Ionisation chambers for radiation protection operate usually in DC mode, with a low-pass filter connected to the anode. The applied voltage V_0 is in the leftmost region in Fig. 3.6, high enough to separate the created electron-ion pairs before they can recombine. The ionisation current comes from the electron-ion pairs created by the secondary electron along its path and is proportional to the energy deposited in the gas by the incident radiation.
- In proportional counters, the field strength $E(r)$ close to the wire is high enough to ionise additional atoms by the secondary electrons. A charge avalanche with an amplification factor $A \approx 10^2 \cdots 10^6$ will result. In a certain voltage range, depending on the gas type and pressure and chamber geometry, the amplification factor A is independent of the energy of the incoming particle, and the thus signal amplitude proportional to particle energy. This is the proportional regime.
- In a Geiger-Müller counter, the electrical field close to the anode wire is so high that the charge avalanche triggers further avalanches all along the wire and depletes the counting gas from neutral atoms. In this regime, the amplification factor is $A \approx 10^{10}$ and independent from the energy of the incoming particle.

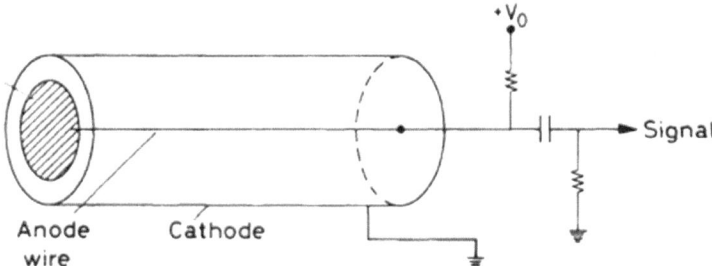

Fig. 3.5 Schematic construction of a gas-filled detector for ionising radiation. (Adapted from [20]. Here, the signal is coupled out with a high-pass filter, in DC ionisation chambers, a low-pass filter would be employed, by changing the position of capacity and resistor to ground)

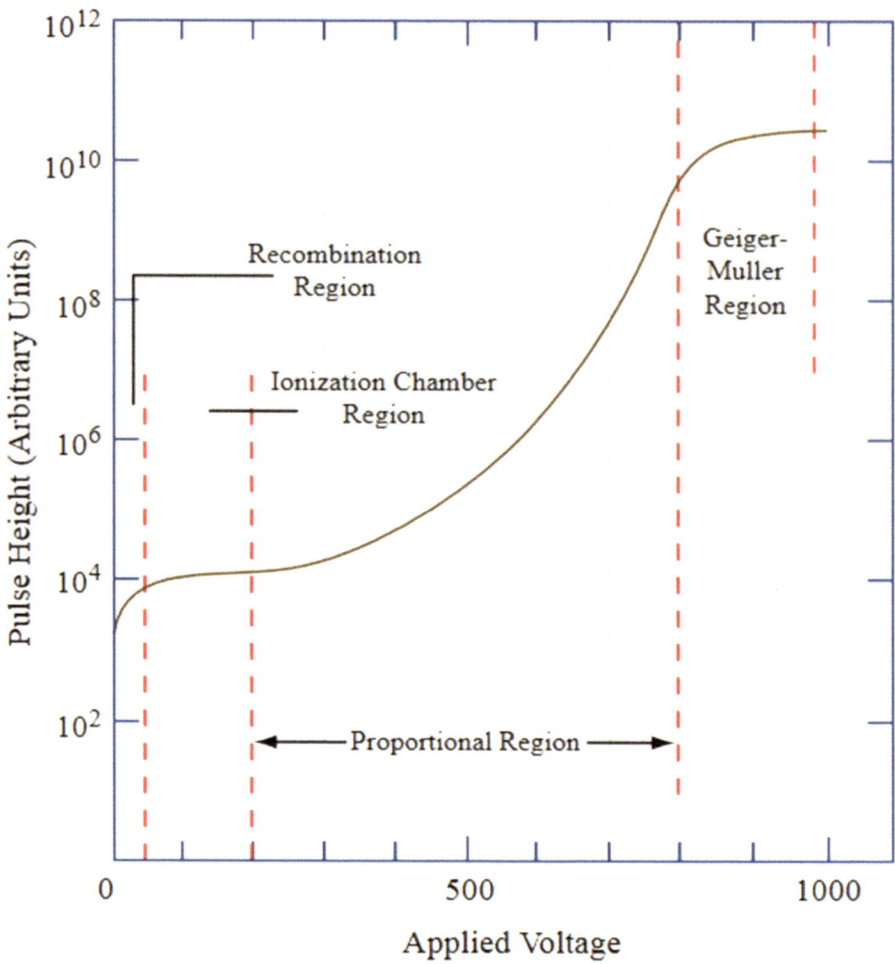

Fig. 3.6 Signal height collected versus applied voltage in gas-filled radiation detectors with central wire. (After [20])

The energy dependence of the three gaseous detector types is mainly determined by the probability that a photon interacts in the wall and emits secondary electrons into the counting gas. Ionisation chambers with walls made from tissue equivalent (TE) plastic and filled with TE gas give a signal proportional to absorbed dose in tissue, which is a good approximation of dose equivalent for photons and electrons. For other particles, a so-called *quality factor* must be applied. In instruments based on Geiger-Müller and proportional counters, metallic filters are placed around the counter chamber to influence the energy dependence of their response to photons. This strategy is successful in a limited energy range, usually between 50 keV and 1.3 MeV or 3 MeV. Below this interval, the dosimeters are insensitive (low energy photons are absorbed ion the detector wall), above they have a strong overresponse.

Proportional counters and Geiger-Müller counters are light and small, their sensitivity is ideally suited to assess photon radiation from activated accelerator components. Ionisation chambers can be used to monitor the radiation emitted by an accelerator in operation. For this, their factory calibration to reference photon sources, must be adapted to take account of the mixed radiation field.

Thermoluminescence and Optically Stimulated Luminescence

Thermoluminescence (TL) and optically stimulated luminescence (OSL) are based on a similar physical effect. Detectors based on this effect are at the basis of most modern passive photon and electron dosimeters. TL and OSL are made from crystalline materials with defined energy bands for electrons, for practical applications they are ground to powder and pressed in tablets or rods (Fig. 3.7). Ionising radiation creates free charges in the crystal's conduction band. Instead of immediately returning to the valence band under emission of light, some electrons are "captured" in so-called trap levels, situated energetically below the conduction band. The trap-levels are metastable states with a long lifetime, and thus capable to store information about the radiation: the number of occupied trap levels is proportional to the dose absorbed in the TL or OSL crystal.

To read out the stored information, one stimulates the electrons in the trap levels either thermally by the application of heat (TL), or optically with a laser of a defined wavelength (OSL). The excitation energy moves electrons from the trap levels to the conduction band, from where they return rapidly to the valence band under emission of light. The photons emitted by the TL or OSL material are amplified by a photomultiplier, the integrated light intensity is proportional to the dose absorbed in the material.

Fig. 3.7 Thermoluminescence material crystal powder, disks and rods of different size, with permission from [21]

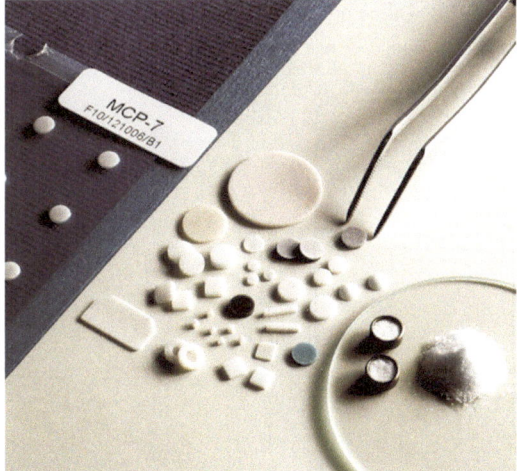

A commonly employed TL material is LiF. Its average atomic number is tissue equivalent. TL detectors from LiF in which the isotope ^{6}Li is enhanced are used as thermal neutron detectors in passive neutron dosimeters (Fig. 3.8).

Semiconductor Detectors

A semiconductor is a crystalline solid from Silicon (Si) or Germanium (Ge), in which the atoms are arranged in a regular, spatially repetitive pattern. The base materials are usually doped with elements in the neighbouring columns of the periodic table of elements. In semiconductor crystals, the highest energy level of the valence band is separated from the conduction band by a gap with a width of a few eV. At room temperature, a few electrons occupy the conduction band and the semiconductor exhibits a small dark current. This can be eliminated by cooling the detector with liquid nitrogen to $T = 77$ K.

The working principle of a semiconductor detector is to release an electron from the valence band and transfer it to the conduction band. The energy required for this process, the *ionisation energy* is composed of the energy to create an electron-hole pair and the energy required to traverse the band gap. It is of the order of a few eV. The hole signifies an empty place in the valence band. The ionisation energy of a semiconductor diode is much smaller than the ionisation energy in a gas-filled detector and for an identical photon energy, more charges are produced in a semiconductor detector than in a gaseous detector. This results in a better energy resolution. The detector also has a higher atomic number (Si: $Z = 14$; Ge: $Z = 32$) and a higher density than gases, these two factors increase the cross section for the photo- and Compton effects and thus the chance to absorb a considerable part of the gamma photon's energy. Such detectors from pure Germanium or, more rarely, pure Silicon

Fig. 3.8 Different TLD holders to be inserted in a filter cassette, providing the correct energy response, with permission from [21]

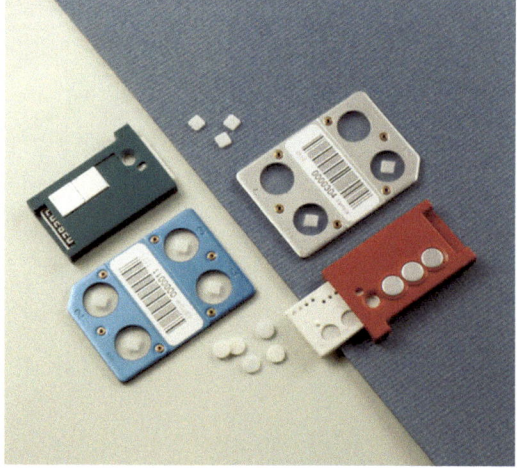

are used in radiation protection as spectrometers for the identification of radionuclides by their characteristic emissions.

For the application as a dosimeter, positively and negatively doped semiconductor layers enclose a layer of intrinsic (undoped) material to form a PIN diode. A voltage is applied across the PIN junction to empty the intrinsic layer from all charge carriers (*reverse bias*). When ionising radiation traverses the intrinsic layer, electron-hole pairs are created and separated by the bias voltage, generating a small charge pulse which is amplified for detection.

The small size of PIN diodes and the possibility to power the bias voltage and the detection circuit with small batteries have made PIN diodes the detectors of choice for personal electronic dosimeters, and many variants of such devices are on the market.

3.4.2.2 Neutron Dosimeters

Like photons, neutrons are indirectly ionising particles. Two physical processes are employed in neutron dosimeters for the detection of thermal neutrons and energetic neutrons.

Thermal Neutron Detectors and Rem-Counters

Thermal neutrons (by convention, neutrons with a kinetic energy of less than 0.025 eV) can be captured by specific nuclei, such as ^{3}He, ^{6}Li, ^{10}B, and ^{113}Cd. After absorbing the thermal neutron, these nuclei become unstable and disintegrate into several easily detectable charged particles. Passive thermal neutron detectors are LiF TL detectors enriched in the isotope ^{6}Li. A second detector of the same size, enriched in ^{7}Li, can be used to estimate the photon component in the radiation field for background subtraction. Active detectors are proportional gas counters filled with either ^{3}He or ^{10}BF$_3$ (a molecule containing boron, and gaseous under normal conditions). Neutrons with higher than thermal energies t must be slowed down in a moderator to be captured with good efficiency in a thermal neutron detector.

Neutron moderators are made from materials which are rich in hydrogen, mostly polyethylene (PE, CH$_2$). In this material, fast neutrons are slowed down in multiple scattering events with protons. In a rem-counter, (from Rem–Roentgen equivalent man, an obsolete unit for an equally obsolete, pre-1990 dose equivalent quantity), the ^{3}He or ^{10}BF$_3$ counter is surrounded by a composite moderator, consisting of PE to moderate and borated plastic to absorb surplus thermal neutrons. (Fig. 3.9). The moderator layers are arranged in a way to model the energy dependent response of the instrument as closely as possible to the fluence-to-dose rate conversion coefficient for neutrons. A rem-counter loses sensitivity for neutrons with energies of more than a few MeV. They are insufficiently moderated by the PE layer and have a small detection cross section in the thermal neutron detector. These rem-counters can nevertheless be used in locations where one has previously performed an in-situ calibration: one determines dose equivalent from high-energy neutrons with an independent method and modifies the calibration coefficient of the standard rem-counter so that is shows the same result. This method requires that the ratio of low-to high-energy neutrons in the radiation field remains approximately independent of the details of accelerator operation, which must be verified independently. As a rule of thumb, a standard rem-counter shows approximately half of the correct dose equivalent when used to verify the shielding of high-energy accelerators.

An independent measurement of the full neutron dose equivalent can be made by a so-called extended range rem-counter. In this instrument, heavy metal layers complement PE and borated plastic. They increase high-energy sensitivity by neutron spallation. The first documented exemplar of an extended range rem-counter can be found in [17], today a few companies manufacture commercial versions of this or similar dosimeters. Extended rem-counters are both heavier and more expensive than standard instruments.

Bonner Sphere Spectrometer

A Bonner sphere spectrometer is a few-channel neutron spectrometer, consisting of up to 15 spherical moderator spheres made from PE, some with metallic inserts to extend the energy range beyond an energy of a few MeV (Fig. 3.10).

Each sphere has a characteristic, energy dependent response curve for neutrons (Fig. 3.11). The count rates for a thermal detector in each of the spheres can be unfolded (a mathematical inversion procedure) to yield an estimate of the energy-dependent neutron fluence spectrum. The dose equivalent can then be determined from the spectrum by multiplying it channel by channel with the corresponding fluence-to-dose equivalent conversion coefficient [25]. The application of a Bonner sphere spectrometer to neutrons from accelerators has been described in [26]. This procedure is time consuming and needs the additional step of spectrum unfolding. It is only employed in research projects when a good knowledge of the spectrum is necessary.

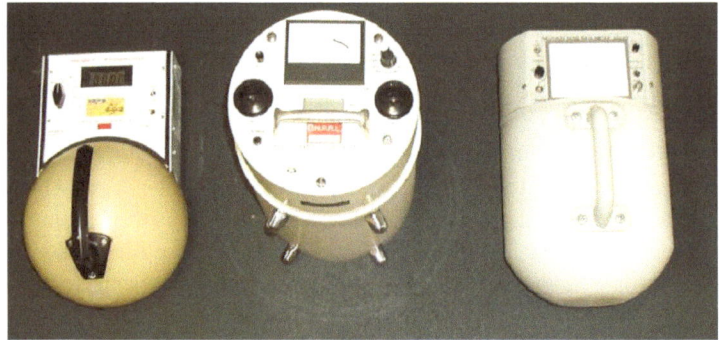

Fig. 3.9 Different commercially available rem-counters. (From [16])

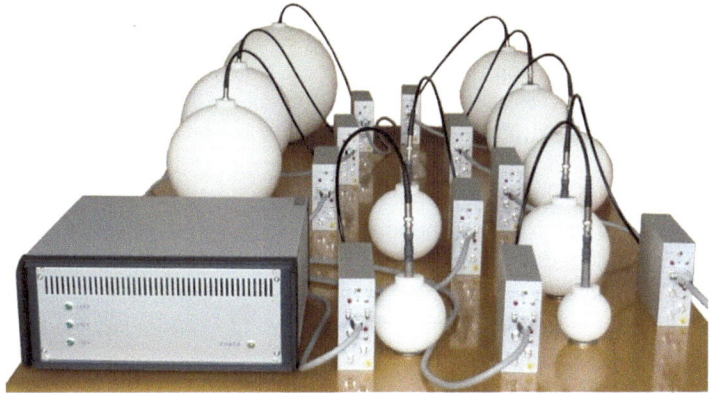

Fig. 3.10 A Bonner-sphere spectrometer consisting of 10 different spheres with amplifiers and data acquisition. [18]

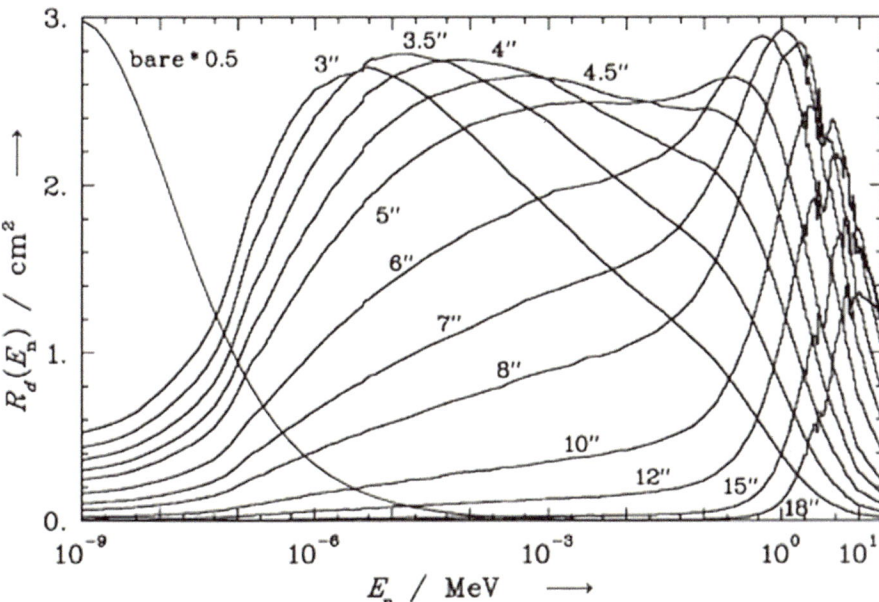

Fig. 3.11 Energy-dependent response functions of spheres constituting a Bonner-sphere neutron spectrometer with 12 different spheres. The diameter of the spheres is given in inch. (1 in = 2.54 cm). (From [25]. Figure reproduced with permission from Elsevier Science & Technology Journals)

Proton Recoil-Based Neutron Detectors

In collisions, energetic neutrons can transfer a large part of their kinetic energy to protons. The recoil proton deposits the kinetic energy by charged particle interactions. Building on this principle, some neutron dosimeters for energies of more than 100 keV consist of a neutron-to-proton converter made from a hydrogen-rich material, combined with a charged particle detector.

An active personal dosimeter combining Si diodes with ^{6}Li converter for thermal neutrons, PE converter for fast neutrons and uncovered diodes for photon detection is described in [27, 28].

As an alternative to a Bonner sphere neutron spectrometer, the proton recoil spectrometer can be employed for neutron energies of more than one MeV [23]. They consist of a proportional counter in which the pulse-height, proportional to the energy of the recoil proton, is registered. After unfolding, the pulse height spectrum yields the energy-dependent neutron fluence spectrum. The advantage over a Bonner sphere spectrometer is, that the pulse height spectrum has more channels, which makes the unfolding process more robust and results in a better energy resolution. In a radiation field where low-energy neutrons make an important contribution to dose equivalent the proton recoil spectrometer can be combined with a few Bonner spheres for energies up to the MeV-range.

The functions of recoil and detection are combined in an ionisation chamber filled with hydrogen gas. This chamber measures the absorbed energy of neutrons by registering their recoil protons released from hydrogen molecules. It is also sensitive to photons and electrons and must be calibrated with a field-specific calibration coefficient. This technique has been employed to measure dose equivalent in radiation fields generated by particle accelerators [24].

3.5   Radiation Protection at Accelerators

At high doses, the exposure to ionising radiation may lead to tissue effects, at low doses and dose rates, stochastic health effects are possible and become more probable with increasing effective dose received. To protect accelerator personnel and members of the public, living in the vicinity of the accelerator, from the detrimental effects of ionising radiation, multiple radiation protection strategies are employed:

First, one strides to reduce one of the main sources of ionising radiation at an accelerator, beam loss. Where this is not possible for physical or operational reasons, beam loss is concentrated at a few locations with the help of collimators (Sect. 2.6.1). These are surrounded with shielding to absorb the generated particle cascades.

The remaining radiation environment at an accelerator consists of radiation penetrating the shielding or released from the accelerator area by way of activated air or cooling water. The largest hazard activated material because many workers are exposed to its decay radiation during repair and maintenance of accelerator components.

3.5.1   Shielding Against Prompt Radiation

In contemporary particle accelerators with high beam energy and intensity, the level of ionising radiation from beam loss during accelerator operation is generally severe enough to present a danger for health and life. Consequently, accelerators are installed either underground, or behind shielding walls from iron, concrete and earth, to dilute and absorb the secondary particle cascades following beam-matter interactions.

Methods and results for the design of accelerator radiation shielding fill numerous scientific publications, summarised in [3, 7, 10, 31, 34]. An estimate of the required shielding can be obtained by estimating the flux of secondary particles emerging from a primary beam collision, to determine the attenuation of this flux by shielding material and to convert it into dose equivalent outside of the shielding. Radiation attenuation models have the general form

$$H(r,\theta) = \frac{1}{r^2} H_0\left(E_p\right) g\left(\theta\right) \exp\left(-\frac{d}{\lambda \sin \theta}\right)$$

Here, H is the expected dose equivalent rate outside of a shielding with a thickness of d under the emission angle θ from the beam loss point or target. H_0 is the energy-dependent emission of secondary particles from the target, conveniently expressed as dose equivalent, and $g(\theta)$ is the angular dependence of the emission. The product $H_0\, g(\theta)$ is called the "source term". λ stands for the attenuation length of radiation in the shielding material. The parameters in the equation depend on the type of accelerated particle because the secondary particle spectra differ strongly between electron-, proton- and heavy-ion accelerators. In Table 3.2, a widely used set of parameters for the shielding of proton accelerators is reproduced from [7].

From Table 3.2 one can see that the attenuation length λ =50 cm for concrete, attenuating a hadronic cascade laterally to the beam direction to 1/e (36%).

Approximate shielding models deliver order-of-magnitude estimates of dose equivalent rates, from which the required thickness of simple concrete and iron shielding walls can be estimated within an uncertainty of one attenuation length. Their simplicity allows to evaluate them with a pocket calculator or a spreadsheet, but they cannot cover more complex accelerator layouts, with openings, access labyrinths, connected galleries and distributed locations of beam loss. For complex shielding arrangements one must resort to Monte-Carlo radiation transport programs such as FLUKA [31, 35] or MCNP [32]. The basic idea behind these programs is to describe the development of the secondary particle cascade, by following the "history" of test particles. At every interaction step, the numerical values of the energy and angle-dependent interaction cross sections are drawn randomly from the respective probability distribution functions. Many thousands or millions of particle histories are followed to yield average results of particle fluence, absorbed dose and activation. By multiplying the estimated fluence with conversion coefficients, one

Table 3.2 Parameters of a frequently used shielding formulas for proton accelerators

	NCRP Report 144 [7]
$H_0(E_p)$ (E_p in GeV)	$2.8\ 10^{-13}\ E_p^{0.76}$ Sv m^2 proton^{-1}
$g(\theta)$ (θ in degrees)	$\exp(-0.040\ \theta)$
λ (g cm^{-2})	$42.8\ A^{1/3}$
concrete (cm)	50
iron (cm)	22

can derive operational radiation protection quantities like dose equivalent. An example of the result of a Monte-Carlo transport calculation for radiation shielding is seen in Figure 2.16.

These simulation programs have been developed over decades by collaborations of scientists and they are regularly validated by comparison with actual dose measurements at existing facilities. Their scope goes beyond the calculation of dose equivalent outside of shielding structures, they also allow the estimation of material damage by accelerator beams, or the dose equivalent from activated materials.

3.5.2   Protection Against Ionising Radiation from Activation

In all types of accelerators, low-energy radiation is emitted in form of photons and electrons by activated accelerator components and structural material. Personnel enters in contact with these radiation fields when accelerator components need maintenance, repair, or have to be exchanged during planned or unplanned shutdowns of the accelerator. In these periods, accelerator personnel accumulate the major part of annual personal doses.

The methods of radiation protection against low-energy radiation are similar to those applied in nuclear facilities, or medical environments: One tries use distance to the source, radiation shielding and limitation of the exposure time to keep the personal doses as low as reasonable achievable (ALARA).

- Establishing sufficient distance to the source of radiation is nearly impossible if the personnel must intervene on activated accelerator material. A possibility for optimisation is the use of teleguided manipulators or robots to accomplish some of the mechanical work. The financial cost and the required additional time make robotic solutions interesting only in those cases where very high dose rates are prevalent, for example on targets and beam dumps.
- Where possible, shielding against photon radiation is employed, in form of lead-equivalent blankets or mobile walls made from heavy metals. This proves often as unpractical, especially when the source of radiation is widely distributed, for example activated, large accelerator components or the walls of the accelerator tunnel.
- Given the difficulties to apply distance and shielding, the most effective method for the optimisation of personal dose during particle accelerator maintenance is

usually the reduction of exposure time. This can be achieved by well-trained personnel, who may have previously rehearsed the operations in the radiation field on non-radioactive mock-ups. Many nuclear establishments have made the experience that the rigorous preparation of work in a radiation field benefits the reliability of the operation, and that the time and money spent for the minimisation of the radiation dose is paid back by an increased efficiency of the intervention.

3.5.3 Control of Radioactive Material

Owners of radioactive material are obliged to exercise close control over the material, to prevent its loss or dissemination in the public. Activated material falls under these regulations as soon as its activity exceeds a legal threshold, the exemption level. If the activity or activity concentration for one isotope exceeds the corresponding exemption level, the material is considered as radioactive. For the European Union, the exemption levels are published in the EURATOM Basic Safety Standards [30]. Radioactive material cannot be traded freely (for example as scrap metal or as filler material in constructions, but it must be stored in intermediate or final radioactive waste depositories. These are operated by organisms authorised by national authorities, and the storage of the material must be paid for.

The transport of radioactive material is also strictly regulated. In Europe, an agreement for the road transport of dangerous goods was concluded in 1957. Its original title in French reads "Accord européen relatif au transport international des marchandises Dangereuses par Route "and it is best known under its acronym "ADR" [29]. Radioactive materials fall in the class 7 of dangerous goods. Depending on the activity concentration, total activity, and dose equivalent rate on the outside of the packaging, shipping of radioactive material must meet several requirements. In the worst case, it can only be transported in special containers by authorised companies.

The price to be paid for the storage and transport of radioactive material make that minimisation of its quantity and its movements is not only a safety consideration, but may also have a positive financial effect.

References

Beam Loss in Accelerators

1. S. Gilardoni, D. Manglunki, *50 Years of the CERN Proton Synchrotron Vol. 1, CERN-20011-04* (CERN, Geneva, 2011). https://doi.org/10.5170/CERN-2011-004

Beam-Matter Interactions

2. A. Bertarelli, *Proceedings of the Joint International Accelerator School: Beam Loss and Accelerator Protection, Newport Beach, United States, 5–14 November 2014, edited by R. Schmidt, CERN-2016-002* (CERN, Geneva, 2016), pp. 159–227. https://doi.org/10.5170/CERN-2016-002.159
3. J. D. Cossairt, Radiation Physics for Personnel and Environmental Protection, FERMILAB Report TM-1834, Revision 15, April 2016, http://esh-docdb.fnal.gov/cgi-bin/ShowDocument?docid=1007
4. E. Iliopoulou et al., Measurements and FLUKA simulations of bismuth and aluminium activation at the CERN Shielding Benchmark Facility (CSBF). Nucl. Inst. Methods Phys. Res. A **885**, 79–85 (2018)
5. D. Kiselev, *Proceedings of the CAS–CERN Accelerator School: High Power Hadron Machines, Bilbao, Spain, 24 May – 2 June 2011, edited by R. Bailey, CERN-2013-001* (CERN, Geneva), pp. 437–463. https://doi.org/10.5170/CERN-2013-001.437
6. A. Lechner, *Proceedings of the CAS–CERN Accelerator School: Beam Injection, Extraction and Transfer, Erice, Italy, 10–19 March 2017, edited by B. Holzer, CERN Yellow Reports: School Proceedings, Vol. 5/2018, CERN-2018-008-SP* (CERN, Geneva, 2018), pp. 47–68. https://doi.org/10.23730/CYRSP-2018-005.47
7. National Council on Radiation Protection and Measurements (NCRP), Radiation Protection for Particle Accelerator Facilities, NCRP Report No. 144, 2003
8. P.A. Zyla et al. (Particle Data Group), to be published in Prog. Theor. Exp. Phys. 2020, 083C01 (2020), http://pdg.lbl.gov/
9. R. Schmidt, *Proceedings of the CAS–CERN Accelerator School: Advanced Accelerator Physics, Trondheim, Norway, 18–29 August 2013, edited by W. Herr, CERN-2014-009* (CERN, Geneva, 2014), pp. 221–243. https://doi.org/10.5170/CERN-2014-009.221
10. W. P. Swanson, Radiological Safety Aspects of the Operation of Electron Linear Accelerators, International Atomic Energy Agency Technical Reports Series No. 188, IAEA, Vienna (1979), https://www-pub.iaea.org/MTCD/Publications/PDF/trs188_web.pdf
11. B. Todd, S. Uznanski, *Proceedings of the CAS–CERN Accelerator School: Power Converters, Baden, Switzerland, 7–14 May 2014, edited by R. Bailey, CERN–2015–003* (CERN, Geneva, 2015), pp. 245–263. https://doi.org/10.5170/CERN-2015-003.245

Ionising Radiation

12. R. Antoni, L. Bourgeois, *Applied Physics of External Radiation Exposure* (Springer International Publishing AG, Cham, 2017)
13. The European Council, Council Directive 92/58/EEC on the minimum requirements for the provision of safety and/or health signs at work, (EEC Brussels, 1992) https://eur-lex.europa.eu/legal-content/EN/TXT/?uri=CELEX:01992L0058-20190726
14. The International Commission on Radiological Protection, The 2007 Recommendations of the International Commission on Radiological Protection, ICRP Publication 103 (2007), http://www.icrp.org/publication.asp?id=ICRP%20Publication%20103
15. The International Commission on Radiological Protection (ICRP) 2011 Statement on tissue reactions (available at www.icrp.org//page.asp?id=123

Radiation Dosimetry at Accelerators

16. R. Tanner et al., Practical Implications of Neutron Survey Instrument Performance, Health Protection Agency (UK) RTeport HPA-RPD-016 (2006), https://www.gov.uk/government/publications/neutron-survey-instrument-performance-practical-implications
17. C. Birattari et al., The Extended Range Neutron rem-counter 'LINUS': Overview and latest developments. Radiat. Prot. Dosim. **76**, 135–148 (1998)
18. ELSE Nuclear S.R.l., Milano. http://www.elsenuclear.com/en/bonner-spheres-spectrometer
19. K. Kleinknecht, *Detectors for Particle Radiation* (Cambridge University Press, Cambridge, 1998)
20. W.R. Leo, *Techniques for Nuclear and particle Physics Experiments*, 2nd edn. (Springer Verlag, Berlin Heidelberg New York, 1994)
21. RadPro International GmbH, Wermelskirchen. https://www.radpro-int.com/tld-1/tld-material/
22. H. Reich (ed.), *Dosimetrie ionisierender Strahlung* (Teubner, Stuttgart, 1990)
23. J. Taforeau et al., Fluence measurement of fast neutron fields with a highly efficient recoil proton telescope using active pixel sensors. Radiat. Prot. Dosim. **161**, 41–45 (2014)
24. C. Theis et al., Characterisation of Ionisation chambers for a mixed radiation field and investigation of their suitability as radiation monitors for the LHC. Radiat. Prot. Dosim. **116**, 170–174 (2005)
25. B. Wiegel, A.V. Alevra, NEMUS – the PTB Neutron Multisphere Spectrometer: Bonner spheres and more. Nucl. Inst. Methods Phys. Res. A **476**, 36–41 (2002)
26. R. Bedogni, Neutron Spectrometry with Bonner Spheres for Area Monitoring in Particle Accelerators, Radiation Protection Dosimetry 146 (2011) pp 383–394
27. M. Wielunski et al., Study of the sensitivity of neutron sensors consisting of a converter plus Si charged-particle detector. Nucl. Inst. Methods Phys. Res. A **517**, 240–253 (2004)
28. M. Wielunski et al., The HMGU combined neutron and photon dosemeter. Radiat. Meas. **101**, 13–21 (2017)

Radiation Protection at Accelerators

29. http://www.unece.org/trans/danger/publi/adr/adr_e.html
30. COUNCIL DIRECTIVE 2013/59/EURATOM of 5 December 2013 laying down basic safety standards for protection against the dangers arising from exposure to ionising radiation (2013)
31. T.T. Bohlen et al., The FLUKA Code: Developments and Challenges for High Energy and Medical Applications, Nucl. Data Sheets **120**, 211–214 (2014). https://fluka.cern/
32. C.J. Werner (editor), "MCNP User's Manual – Code Version 6.2", LA-UR-17-29981 (2017)
33. A.H. Sullivan, *A Guide to Radiation and Radioactivity Levels near High Energy Particle Accelerators* (Nuclear Technology Publishing, Ashford, 1992)
34. R.H. Thomas, G.R. Stevenson, *Radiological Safety Aspects of the Operation of Proton Accelerators*, International Atomic Energy Agency Technical Reports Series No. 183 (IAEA, Vienna, 1988)
35. G. Battistoni et al., Overview of the FLUKA code, Ann. Nucl. Energy **82**, 10–18 (2015)

Chapter 4
Industrial Safety at Particle Accelerators

Abstract The construction and operation of particle accelerators implies the use of numerous technologies and trades which are well-known from the manufacturing and construction industries. Consequently, their safety hazards are described in the literature and standard best practice solutions exist for controlling the risks emerging from these activities. In this section, the occupational hazards of electricity, mechanical equipment and pressure vessels are illustrated with examples from particle accelerator facilities. Further sections are dedicated to accelerator-specific protection against fire, occupational noise, and environmental damage.

4.1 Electrical Safety

The fundamental principle of accelerator operation is the movement of charged particles by electro-magnetic forces. Electrical current generates magnetic fields and high frequency electrical power drives RF cavities, the two technologies at the heart of every accelerator. Electrical hazards from accelerator equipment are described in the relevant sections (Safety Aspects of Magnets, 2.2.3 and Hazards from RF systems, 2.4.3).

Electrical energy is also often the most convenient form of transmitting and using energy. It is found wherever objects must be moved, or matter converted. This justifies highlighting electrical safety in this section from a general point of view (Fig. 4.1).

4.1.1 Electrical Hazards

If electrical energy is released in an uncontrolled way, different types of harm may result. One or more of the following electrical hazards may occur:

T. Otto, *Safety for Particle Accelerators*, Particle Acceleration and Detection,
https://doi.org/10.1007/978-3-030-57031-6_4

Fig. 4.1 Warning sign
against electrical risk, after
[1]. (Image source: https://
publicdomainvectors.org)

Table 4.1 Physiological effects of electrical current traversing the body

Current flowing through the body (mA)	Physiological effect
0.5–2	Sensation threshold for most individuals
2–10	Tingling sensation, muscle tremor, onset of pain
10–60	Muscle contractions, inability to let go, breathing difficulties
>60	Ventricular fibrillation, cardiac arrest, extreme muscle contraction, burns at contact points and in deep tissues

4.1.1.1   Electric Shock and Burns

Electric shock may arise from *direct contact* with current-carrying parts, for example when a person touches a live conductor that has become exposed because of damage to the insulation of an electric cable. Alternatively, it may result from *indirect contact* if, for example, an internal fault results in the exposed metalwork of an electrical appliance, or even other metalwork such as a sink or plumbing system, becoming live.

In either case there is a risk of an electric current flowing to earth through the body of a person who touches the live conductor or live metalwork. The severity of the electrical shock is determined by electrical voltage and current. The resistance of the body depends on the applied voltage and on environmental factors, foremost humidity. As a rule of thumb, dry skin isolates against voltages below 50 V. Higher voltages can overcome higher electrical resistance, for example of insulating tool handles, gloves, and at very high voltages even the resistance of air. Once the body connects electrical current to ground, the magnitude of electrical current flowing through the body determines the extent of the damage (Table 4.1).

For voltages above a few hundred volts, the resistance across the body between hand and foot or between both hands amounts to about 1 kΩ. Connecting the standard voltage of 230 V to ground could therefore lead to a current of 230 mA traversing the body. Accidental contacts with the 230 V mains are not always deadly because the accident victims may have a higher-than-average skin resistance, carry shoes with insulating soles, or their unconscious reflex of pulling back the hand is fast and strong enough to safe their live.

As a consequence of the muscular contractions of a non-lethal electrical shock, the victim may be thrown off balance and fall, for example from a ladder, with severe consequences. Non-lethal electrical shocks can also lead to potentially deadly heart arrhythmia.

4.1.1.2 Fire

Electrical current leads to heating of ohmic resistances or may cause sparks at badly executed electrical connections. Both phenomena can ignite surrounding material and lead to a fire. Approximately 25% of domestic and industrial fires have their origin in electrical malfunctions, for example faulty equipment or overloaded circuits. An additional hazard exists in environments where flammable liquids or gases are stored and where excessive heat or sparks can lead to an explosion. Finally, an electrical arc (see next paragraph) may ignite a fire.

4.1.1.3 Electrical Arc

Multiple causes can provoke an electrical arc (arc flash):

- At high voltage, an electrical spark can jump over an air gap. The distance it can traverse increases with the applied voltage. The initial spark ionises air molecules along its path which become conducting for a short amount of time.
- A short circuit between two conductors, for example by connecting them accidentally by a metallic object, suddenly lowers the electrical resistance to a very small value and allows a high current to flow, which can melt the object causing the short-circuit.
- Condensation or water projection can cause the breakdown of electrical insulations, making a path for electricity to connect two conductors or to ground.

In an electrical arc, a large amount of energy is released in a short time interval. The temperature in the arc may attain several thousand degrees, leading to the emission of a broad spectrum of electromagnetic radiation between ultraviolet and heat radiation, and a sudden expansion of the surrounding air, causing a shock wave.

The consequences for a person struck by an electrical arc are similar to being struck by a lightning (a naturally occurring electrical arc) and potentially lethal: electrical shock, burns by the electrical current and radiant heat, exposure to UV radiation and noise, impact of projected high-speed hot or molten fragments.

4.1.2 Electrical Safety

4.1.2.1 Electrical Conformity

One aspect of electrical safety is the sound design and construction of electrical devices. Correctly dimensioned electrical installations minimise the risk of electrical faults during standard operation and provide protection to users if faults should nevertheless occur. In the European Union, the Low Voltage Directive 2014/35/EU [3] sets the frame for conformity of electrical equipment with operating voltages of up to 1000 V (AC) or 1500 V (DC), but, in contrast to other directives, the "Principal safety objectives for electrical equipment" in Annex I are not detailed prescriptions but only a declaration of broad goals. One of the declared aims of the European directive is to protect persons of harm from electrical hazards if the electrical equipment is used according to its intended purpose.

Detailed technical requirements for the construction and testing of electrical equipment are found in the harmonised standards to the directive, listed by the European Commission [2]. This document is regularly updated and contains references to approximately 700 standards published by CEN and CENELEC (see Annex B) regulating all aspects of design and construction of low-voltage electrical equipment for use in the European Union. Should no harmonised standard exist for a specific safety aspect, then the Directive states that the relevant standards of the International Electrotechnical Commission shall be used. An important prerequisite for obtaining a conformity certificate is a complete and accessible documentation of the equipment, not only for its use but also for repair and maintenance.

The declaration of conformity with the Directive and its harmonised standards is a self-declaration by the manufacturer, who engages his full responsibility. Certification by an independent testing laboratory is not required for the declaration, but it is advisable to consult specialists from electrical testing laboratories on the relevance and application of the various standards. This applies also to electrical devices which are purpose-built in one or a few exemplars for specific needs in a particle accelerator, and which cannot be found by a commercial supplier.

High voltage equipment is not covered by a European directive but standards for such equipment are published by IEC and CENELEC.

Standards for electrical equipment with a global scope are published by the International Electrotechnical Commission IEC.

4.1.2.2 Practical Electrical Safety

The consequences of an electrical accident can be severe, even lethal, and suitable means for protection against the electrical hazard must be taken. Electrical equipment and assemblies meeting the requirements of conformity do not present electrical hazards when they are used in accordance with manufacturers directives. Critical phases in the life cycle of electrical equipment are maintenance, repair, and modification. In these phases, protections are removed from the equipment and electrical conductors become accessible for direct contact.

Workers engaged in maintenance, repair and modification of electrical equipment must have a *professional competence* in electricity or electronics and specific knowledge of the equipment that they are asked to work on. The professional competence is obtained during vocational training, higher education, and courses in the frame of continuous education. In France it must be periodically complemented by specific, electrical safety-oriented training courses as required by law [4], which singles out the French approach within the European Union. *Equipment-specific knowledge* is gained from user- and maintenance manuals, complemented by electrical circuit diagrams. This underlines the importance of proper documentation in the process for obtaining conformity.

Before engaging work on electrical equipment, workers must make certain that it is no longer connected to the supply voltage, and that it cannot be reconnected during the work. For small equipment this is simply achieved by pulling the power cord. For installed equipment supplied by fixed cables coming directly from a transformer, a process called *lockout* is applied. This process is an example of a *safe system of work* (Sect. 5.2.2).

The lock-out process follows five steps which must be executed in order, and without omission, to create a safe working environment:

1. **Identification of the connection:** the worker must identify the output of the transformer or a switchboard which powers the equipment under question. This is very important in large installations where multiple equipment is connected via switchboards to the same source of electrical power.
2. **Separation from the source:** the identified connection is separated from the power source. This separation is secured by the worker, for example with a personal padlock. This step gives the whole procedure its name, *lockout*.
3. The operator leaves a signed note that (s)he has performed the lockout, this is called **tag-out**.
4. Then, the operator **verifies on the equipment that it is indeed disconnected** from the electrical power source and that all capacitors and impedances, able to store energy, are short-circuited. He uses standard electrical measurement instruments for this.
5. Finally, the operator **connects the equipment safely to ground**, making all attempts of repowering impossible.

In its simplest realisation a single worker is in charge of the whole lockout process. It is unrealistic to assume this to be possible in all but the smallest accelerator facilities. Usually, the responsibility for the components of the accelerator chain is shared between different entities, and the power sources may be geographically far from the equipment fed by them. These facts complicate the simple lock-out process with padlocks, as in the following example:

Before a normal-conducting magnet is maintained, it must be brought in a safe state by disconnecting it from its power source. The magnet expert requests from the electrical power group the separation of the specific magnet from the current source. A member of the electrical power group will separate the connection feeding the indicated magnet from the current source. He may secure the separation with his personal lock. He will inform the magnet expert of the separation. Multiple errors may occur at this stage:

- the electrical expert misreads the identification of the requested magnet or he accidentally locks out a different magnet.
- the electrical expert reconnects the magnet before the work on it has terminated

Note that in a process with distributed responsibilities the magnet expert cannot apply a personal padlock to secure the separation of the power source. In general, he lacks the competence to identify the power sources.

This makes the independent verification of the absence of electrical power and the connection the magnet to ground essential and virtually life-saving for the magnet expert.

The evolution of the lock-out process to complex scenarios with distributed responsibilities is treated in a general way in Sect. 5.2.2 under the headline "Safe Systems of Work".

The best performing system of electrical lockout cannot prevent accidents in the cases where workers must approach electrical equipment while it is powered. These workers must be equipped with personal protective equipment (PPE) protecting against electrocution and arc flash (Fig. 4.2):

- Insulating gloves, with appropriate voltage rating. These gloves may have a short shelf-life and must be exchanged regularly.
- Safety shoes with an isolating sole
- A helmet with a face-shield, both resistant to heat and to projected fragments during an arc flash
- A jacket or shirt from non-ignitable fabric

Fig. 4.2 Electrician equipped with PPE: Helmet, arc-flash resistant visor, insulating gloves, fire-resistant jacket. Not visible: shoes with isolating soles. (Image credit: SUVA)

4.2   Mechanical Safety

Mechanical hazards occur when masses are moved or altered. The functional elements of an accelerator, often with a mass of many tons, are produced and maintained in mechanical workshops, must be moved into place by cranes and transport vehicles and connected with each other electrically and mechanically. This section starts with a view on machines and then focusses on transport hazards.

4.2.1   Machines at Particle Accelerators

Machines, defined as mechanisms animated by other than human forces [5], are ubiquitous in a particle accelerator during all phases of the life cycle:

- Production: an accelerator centre does not have the capacity to produce the elements for kilometre-long facilities, and the components for large particle accelerators are built in industry. The centre will still have workshops to produce prototypes or small series of specialised items. These workshops are equipped with standard machine tools (milling-, drilling- and turning machines). Special machines have functions like winding magnet coils, polymerising them and assembling whole magnets in mechanical presses. These machines are tailor-made for purpose.
- Operation: some elements of a particle accelerator are moved by motors or actuators and constitute "machines" according to the strict definition above: vacuum valves, beam intercepting devices and elements of beam instrumentation. For the next generation of particle accelerators with very small beam sizes, remotely operated, motorised systems for aligning the accelerator components are under discussion.
- Maintenance: the maintenance of accelerator elements is sometimes performed in-situ with portable machines, more extensive revisions are done in workshops equipped to disassemble and re-assemble magnets, cryostats, and other equipment.
- Decommissioning: accelerator components which have failed or have reached the end of their lifetime are disposed of. The separation of the radioactive components must be sorted by activity level and by nature of the material, requiring mechanical tools. Certain waste categories can be compacted with help of mechanical presses.

Warning signs symbolize machine hazards, such as hand injury, entrainment and crushing (Fig 4.3).

Fig. 4.3 Warning signs against hazards from machines: hand injury, entrainment, crushing, after [1]. (Image source: https://publicdomainvectors.org)

Fig. 4.4 Mechanical hazard triangle. The adverse interaction of a mechanical element its energy and the operator may lead to an accident. After [11]

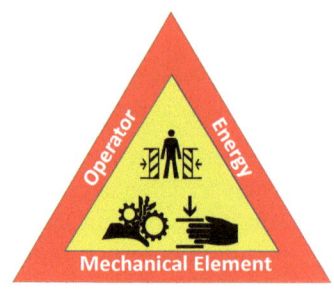

4.2.2   Machine Safety

Most of the time, the function of a machine can be reduced to the movement of masses: a rotating drill, the cutting tool of a milling machine, the movement of the ram in a mechanical press require forces which may cause injury when directed against a person. This is expressed in the mechanical hazard triangle (Fig. 4.4).

Extensive guidance exists for the control of mechanical hazards in the manufacturing industry, covering the use of all types of machine tools and of hand-held power tools [8, 11].

4.2.2.1   The European Machinery Directive

In the European Union, safety aspects of machinery are regulated in the Directive 2006/42/EC of 17 May 2006 on machinery [5, 6]. Under EU regulations, the content of directives must be adopted by the member states into national law within a reasonable delay. The purpose of the directives with focus on safety is twofold: on one hand, all EU members states shall apply the same safety standards on products, so that manufacturers can sell them unhindered in the territory of the EU. On the other hand, the directives are an instrument to guarantee minimal safety standards for the workers and the public for products imported from countries outside the EU.

The directive lists in its Annex 1 the essential health and safety requirements that machinery must meet before it can be introduced to the European Market. The 2006 edition of the directive also counts the construction of machines for one's own use as introduction to the market, so that these devices must meet the same standards as goods for sale. This is obviously in the interest of protecting health and safety of the workers and operators. A machine conforming to the standard may bear the CE-marking, described in detail in Annex III of the directive and reproduced in Fig. 4.5.

A voluminous application guide to the Machinery Directive is available from the European Commission [6] as well as numerous titles of secondary literature, for example [12].

4.2.2.2   Conformity with the EU Directive

The straightforward way to obtain conformity with the European Directive is to construct a machine following a *harmonised standard*. Machinery constructed under respect of the technical requirements in a harmonised standard is presumed to conform to the legally binding European Directive. The list of harmonised standards is published by the European Commission [7].

In the European Union, it is advisable to buy standard machines from established manufacturers who will deliver a CE-certified machine, either from a EU member state or from countries with a long-standing commercial relationship with the EU.

The demonstration of conformity with the European Directive is more time-consuming and possibly costly for modified, entirely self-built, or imported machines. In international research centres, such as particle accelerator facilities, it is common that collaborators make in-kind contributions. These may come in form of a machine in the broad definition of the EU, and then conformity with the European Directive must be sought. A starting point for a self-built machine would be to follow a harmonised standard for conception and construction. Foreign machines or modified, old machines one must be document in a technical file, as

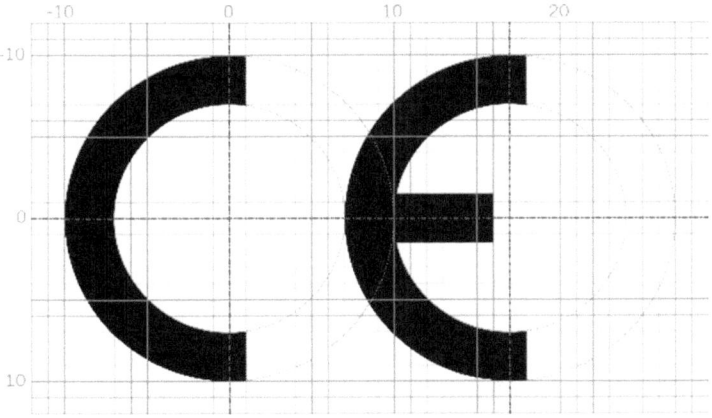

Fig. 4.5 CE-Marking, after [5]. (Image source: https://publicdomainvectors.org)

described in Annex VII of the Directive. This process is called "assessment with internal checks" and is generally sufficient to attest the conformity of a "machine" installed in a particle accelerator, such as movable collimators or elements of beam instrumentation.

The prescriptions of the directive are not mandatory for machinery specially designed and constructed for research purposes for temporary use in laboratories (Article 1.2(h) of the Directive). This leaves a certain margin for very specific, self-built machinery in particle accelerators. In case this provision shall be used, it is advisable that a complete technical documentation, user manual and a risk assessment of the equipment is drawn up be able to mitigate any obvious safety risks. The risk assessment can take account of workers' training in the use of a specific machine.

These remarks are valid, in slightly modified form, for obtaining the conformity for other categories of regulated products in the European Union (Annex B).

4.2.3 Transport at Particle Accelerators

Depending on the size of a particle accelerator, transport of persons and goods is an activity consuming time and resources. Warning signs for transport hazards are illustrated in (Fig. 4.6). The presently largest particle accelerator in the world, CERN's Large Hadron Collider, has a circumference of 27 km. It touches the main site of the organisation in Meyrin (Canton of Geneva, Switzerland) at one point, from where the farthest point is at 8.6 km as the bird flies. Depending on traffic density, the transfer by car may take between 20 and 30 min between the access points on the perimeter. Personnel and material are transported on public roads with conventional vehicles. Accelerator magnets, 15 m long and with a mass of 15 tons, are lowered by overhead cranes into a tunnel connecting to the LHC and are moved on a special magnet transporter (Fig. 4.7). The largest distance to be covered undereground is 30 km. The transporter advances automatically along the tunnel with an optical guidance system at a speed of 3 km/h, an operator is on board for emergency action.

Overhead cranes play an important role in transport for a particle accelerator: heavy, bulky items are lowered into the underground areas with cranes, and they serve to move equipment in the accelerators (Fig. 4.8), in workshops and in testing areas.

4.2.4 Safety of Transport and Handling

As in machines, mechanical forces are employed to transport objects. These forces and the inertia of the loads are increasing with the mass and the speed of the moved objects. Figures 4.7 and 4.8 show typical transport situations in a large accelerator, they have a high potential for accidents if not handled correctly.

Fig. 4.6 Warning signs against transport hazards: overhead load, industrial vehicle, after [1]. (Image source: https://publicdomainvectors.org)

Fig. 4.7 Special vehicle for the transport of superconducting magnets in CERN's LHC tunnel. The vehicle draws electrical energy by the orange catenary from an overhead rail. The vehicle advances by an optical guidance system, following the white line painted on the floor. The white structure to the left is the cryogenic supply line for the superconducting magnets. Copyright CERN, reused with permission

For road transport, one can rely on vans and lorries driven by professional, certified drivers. National regulations define minimal safety standards for road vehicles, they must be checked periodically by certified bodies. All lifting gear (cranes, chains, ropes, slings, hooks) must be regularly inspected for signs of ageing.

In the European Union, special transport vehicles, such as those operating in accelerator tunnels, fall under the Machinery directive [5], with additional requirements in Chap. 3 of Annex I.

Fig. 4.8 Handling of an LHC dipole magnet with an overhead crane in the LHC tunnel. Copyright CERN, reused with permission

The employer is responsible to supply conforming and tested transport equipment to his employees. To prevent transport accidents, a few core rules should be applied:

- Transport operators and vehicle drivers must be trained in the use of their equipment, including the safety aspects. Under certain national legislations, professional certificates are required to use mechanical transport and lifting equipment, elsewhere an instruction by the employer is sufficient.
- Operators and drivers must be well rested and concentrated before their shift.
- Transport and lifting operations shall be scheduled at moments where no other activities take place in the area. This avoids putting bystanders in danger and gives the operators the liberty to choose the optimal path of transport. For example, the magnet transport through CERN's LHC tunnel (Fig. 4.7) takes place from 17:00 h on, after normal working hours.
- The transport/lifting equipment must be suitable for the task. Questions to be asked before every use are: is the capacity sufficient for the masses to be moved? Is it in good shape, no deformation or excessive traces of wear and tear?

Lifting and transport operators shall wear personal protective equipment:

- Safety shoes, to prevent slipping and to provide a limited protection of the toes against heavy loads.
- A safety helmet, in Europe according to standard EN 397, to prevent head injury from small pieces falling off from the load.
- Handling gloves, to protect the hands from bruises, cuts and abrasions while guiding and directing the load.
- High-visibility clothing, to provide optimal visibility by the other operators.
- If necessary, this equipment must be complemented by hearing protection or respiratory protection.

Guidelines for safe transport at the workplace can be found for example in [9].

4.2.4.1 Manual Handling

Despite being at the forefront of science and technology, and of having a parc of sophisticated transport and handling devices, manual handling is sometimes unavoidable at particle accelerators. On the last metre of a transport path, where an equipment must be hauled into its position, manual handling is often the last resort, for lack of adapted mechanical means or for lack of space for applying them. Manual handling bears the risk of injury from sharp edges, rough surfaces and from heavy loads getting out of control from the handlers. The risk of occupational illness such as back pain and other musculoskeletal disorders (MSD) is high, and these illnesses can become permanently disabling. Manual handling of heavy loads more than 25 kg shall be avoided wherever possible, mechanical handling aids shall be used, and only when this is impossible, the load must be shared among several workers.

Manual handling guidelines, especially with respect to avoiding occupational illness are given in [10]. In addition to authoritative information, this website provides tools for handling risk assessment.

4.3 Pressure Vessels

Recipients containing fluids (liquids or gases) at a pressure higher than atmospheric pressure are termed *pressure vessels*. The mechanical energy stored in the pressurised fluid is equal to the product of pressure and volume:

$$W = p \cdot V \left[\mathrm{Pam}^3 \right] = \left[\mathrm{J} \right]$$

If this energy is released instantaneously, severe damage to personnel and equipment may result. In addition, the hazards of the pressurised fluids must be considered, for example cryogenic, flammable, or toxic fluids (Fig. 4.9).

Fig. 4.9 Warning sign: Bottle under pressure, after [1]. (Image source: https://publicdomainvectors.org)

4.3.1 Pressure Vessels at Accelerators

A typical field of application for pressure vessels at particle accelerators is as a part of cryogenic systems (see Sect. 2.3). They are used for containing the cooling medium of superconducting devices, and to contain cryogenic liquid particle detectors (Argon, Krypton and Xenon calorimeters). Gases and fluids under pressure are used for cooling detectors at temperatures close to 0 °C. Conventional applications of pressurised fluids include pressurised air, sometimes used in workshops as a power-transmitter for driving tools without electricity. Pressurised air is also used to move actuators, for example in vacuum valves.

The components for conventional applications can be bought from manufacturers who build the equipment according to legal prescription, for example to the European Directive and harmonised standards in the EU. The institutions developing superconducting or particle detector systems with pressure vessels are responsible for meeting the applicable legislation and standards.

4.3.2 Pressure Vessel Safety

Pressure vessels are ubiquitous in the processing industry and they may present a considerable accident risk with severe consequences [39]. The rupture of a pressure vessel results in the following accident hazards:

- Release of the fluid at high pressure by the leakage, leading to mechanical injury.
- Other than the risk from a pressure jet, the released fluid may be flammable or have chemical or cryogenic hazards, exposing workers and bystanders to additional risks after the release has terminated.
- Disintegration of the whole vessel by explosion, followed by projection of fragments and pressurised fluid.

Badly designed pressure vessels can be the source of severe accidents leading to loss of property and possibly of life.

Two principles in pressure vessel design protect against accidental rupture and release of the contents:

- First, a solid construction of the vessel following current engineering practice and described in national or international standards. The mechanical resistance of the construction materials (usually steel, sometimes other metals for special applications) and their assembly balances the expected pressure variations of the contents up to a maximal operational pressure or service pressure P_s.
- Second, for the case that pressure in the vessel rises accidentally to values higher than P_s, *pressure relief devices* are foreseen to release some fluid and thus lower the pressure. These valves or burst-discs open at a predefined set-pressure and vent the contents of the vessel in a controlled way. Pressure relief devices are mounted in pairs, a relief valve which opens at a pressure P_{v1}, is combined with a burst-disc with set-pressure $P_{v2} > P_{v1}$. The burst-disc opens in two failure cases:
  - the relief-valve fails mechanically, or
  - the sudden mass flow to be evacuated surpasses the valve's capacity.

It is a single-use device and must be replaced after opening. The pressure relief devices must have a sufficiently large opening surface to release the overpressure. The calculation of safety devices is treated in international standards for ordinary [14] and for cryogenic pressure vessels [13]. For cryogenic pressure relief devices, the calculations consider the heat influx to the cryostat, the temperature-dependent thermodynamic parameters of the fluid and phase changes from liquid to gaseous. In the regime where the operator of the pressure vessels works with a notified body, the existence and performance of safety devices becomes an important part of the safety assessment.

4.3.3   The European Directive on Pressure Vessels

Legislation in all countries takes account of the dangers of pressure vessels by prescribing minimal safety standards. In member states of the European Union, the Pressure Vessel Directive [40] regulates the manufacturing and use of pressure vessels. The deadline for transferring its principles into national law in the member states was mid-2016. With this directive, the European Commission aims to harmonize safety standards in the Union, so that goods containing pressure vessels can be freely exchanged on the internal market while at the same time minimal safety standards for workers and consumers are respected.

Annex 1 of the directive lists the essential safety requirements (ESR) which must be met by pressure vessels which are placed on the market, i.e. made available to customers. For inert gases, as helium, the ESR are mandatory if the service pressure PS of the vessel exceeds 1000 bar (the directive still quotes pressures in bar, as in engineering practice) or the volume V exceeds 1 litre (L) and the product PS V is larger than 50 bar L.

Then, vessels are subdivided in those carrying dangerous fluids (for example flammable, oxidising, or acutely toxic gases and liquids), generally not of concern at accelerators, and those for other fluids. They are further classified in 4 categories by the product of service pressure PS and volume V. The categories determine the procedure according to which the manufacturer must demonstrate conformity with the ESR. For vessels in the lowest category I (PS V < 200 bar L), the mechanism of "internal production control" is deemed sufficient to prove that the ESR are fulfilled. This requires the drawing up of detailed technical documentation and a risk assessment. During production, internal controls shall be made to ascertain that the technical documentation is followed.

From category III onwards (PS V > 1000 bar L), a *notified body* must be consulted during the conformity assessment process. Notified bodies are entities accredited by a Member State of the EU to assess whether a product meets certain standards. Familiar examples are the associations or companies entitled to periodically check the traffic-worthiness of a car. As for machines (see Sect 4.2.2), the construction of pressure vessels for proper use falls under the application of the directive [40]. Consequently, the consultation of a notified body must be considered in cost- and schedule estimates for the construction of pressure vessels of categories III or IV for an accelerator facility.

4.4 Fire Safety

Fire hazard is found at workplaces and in homes likewise. Accelerator facilities and their associated workshops, test areas and offices make no difference, and fire prevention is an important point in a safety prevention programme. In this section, the conditions for a fire to break out are enumerated, the potentially devastating effects of a fire described, and prevention measures derived (Fig. 4.10).

4.4.1 The Fire Triangle

Three ingredients are necessary to start a fire and make it grow, they are often represented in form of the "Fire Triangle" (Fig. 4.11):

Fig. 4.10 Warning signs against fire hazards: flammable substance, explosive substance, oxidant substance, after [1]. (Image source: https://publicdomainvectors.org)

Fig. 4.11 The fire triangle: The combination of fuel, oxygen (or another oxidizing substance) and heat, or more general an ignition source may lead to the outbreak of a fire

1. Fuel: burnable substances are fuel for a fire. They can be solids (for example wood, paper, cardboard, packing materials, plastics), liquids (for example solvents, paint and varnish, petrol) or gases (for example butane, propane, acetylene, hydrogen).
2. Oxygen: from the chemical point of view, at the heart of a fire are exothermal oxidation reactions, therefore no fire without oxygen. The oxygen content in air is often enough for keeping a fire going, it may be enhanced by a ventilation system, carrying fresh air in a burning area, or by oxygen in cylinders for welding or for experimental purposes.
3. Ignition source: the presence of fuel and oxygen is not enough to ignite a fire; an ignition source is necessary to provide the energy to start the exothermic reaction. Ignition sources may be naked flames, or smouldering ashes (from smoking materials, in laboratories), sparks (from grinding, welding and faulty electrical equipment or static electricity), hot surfaces (from insufficiently ventilated or cooled electrical and mechanical equipment, lighting, heating).

A fire may start when an ignition source comes close to fuel in a well-ventilated atmosphere. Easy flammable fuels (characterised by a low temperature of ignition) often start the fire, which can spread to other, less flammable materials. Once started, a fire will continue to burn and spread as long as fuel and oxygen are available.

4.4.2   Fire Hazards at Accelerators

A fire hazard exists at those locations, where the three elements of the Fire Triangle (Sect. 4.4.1) come together. Oxygen in air is ubiquitous, and it is enough to consider ignition sources and flammable materials (fuel).

The availability of ignition sources for a fire depends on the operational state of the accelerator:

- During accelerator operation, all electrical systems are powered. Electrical resistances, short circuits or sparks can generate enough heat to set neighbouring material at fire.
- During periods where the accelerator is stopped, many of the accidental ignition sources from electricity are turned off. In these phases the focus is on work

activities using open flames, hot points or producing sparks, like welding, brazing, and grinding. When fire prevention measures during this work are neglected, flammable materials close to the work site may be ignited.

Various types of flammable material can be found in an accelerator facility:

- Cable insulations and other organic materials may burn readily once lit up. An electrical fire can start by igniting a small quantity of insulating organic material and then spread over to the contents of an electrical rack. Cable insulations are often rated for fire resistance, but with enough heat energy from a fire close by, they will eventually start to burn.
- Recently, metal-ion electrical storage batteries are gaining importance, for example Li-ion batteries. They consist of Lithium- and carbon electrodes, separated by an electrolyte made from an organic, flammable substance. Short circuits in these batteries may ignite the electrolyte and lead to a runaway reaction with very high flame temperatures from the oxidation of the alkali metal. Li-ion battery fires in electrical vehicles have been reported to destroy the battery and igniting flammable material in its vicinity [19].
- Another source of fuel is packing material, like carton and wood. Packing often contains fillers made from polystyrene, which is easy to ignite. These materials are brought to the accelerator area with new equipment, and they may accumulate if no regular housekeeping is performed.
- Some detectors for particle physics use flammable gases in ionising radiation detectors, or flammable solvents for certain types of scintillation detectors.
- Flammable solvents are also found in laboratories with frequency-tuneable dye lasers, and in workshops, where they are used for many applications.

4.4.3   Tunnel Fires

Tunnel fires came into the light of public attention with the tragic events in the Mont-Blanc road tunnel between France and Italy in 1999, the Kitzsteinhorn funicular, Austria in 2000, and the underground metro station fire in Daegu, South Korea in 2003. The enclosed nature of tunnels gives rise to special fire phenomena in conjunction with the ventilation. Escape of personnel and access for emergency forces is more perilous and can become impossible if temperatures rise too high or smoke obstructs the pathways. A comprehensive reference for tunnel fires is [15].

The potential causes of fire in accelerator tunnels are different from road or rail tunnels, but the dynamics of fire development and fire spread, and its consequences are similar.

4.4.3.1   Fire Dynamics in Tunnels

Fires in open air radiate heat in all directions and receive a nearly unlimited oxygen supply. In compartments (closed environments), the walls reflect a part of the fires heat energy and lead to more intensive burning. Also, the supply of oxygen is limited. One commonly used parameter to describe the size of a fire is the *Heat Release Rate* (HRR).

One speaks of a *fuel-controlled* or oxygen-rich fire when the rate of combustion and the HRR are dominated by the availability of combustible materials. Oxygen is available in sufficient quantity. In fuel-controlled conditions, the HRR of the fire grows until all fuel is consumed.

In a *ventilation-controlled* or oxygen-starved fire, there is not enough oxygen to burn all the fuel available. Oxygen supply is limited by the enclosure of the fire zone. The heat of the fire releases flammable gases from the fuel, but these are incompletely combusted. In ventilation-controlled conditions, the HRR of the fire stagnates. The air flowing out of the fire zone is entirely depleted of oxygen and loaded with toxic carbon monoxide and partially combusted fire gases. If a ventilation-controlled fire is suddenly supplied with oxygen (by opening a door or window in a compartment, or by activating mechanical ventilation in a tunnel), the hot, unburned fire gases ignite spontaneously and lead to the phenomenon of "backdraft", a sudden development of fire engulfing a large area, radiating a lot of heat, and dangerous for evacuating persons and fire fighters.

Ventilation-controlled fires are a concern in compartments such as electrical substations or chemical storages. Generally, in tunnels, the oxygen supply is high enough for a full combustion of the fuel. The onset of a ventilation-controlled fire can be estimated by putting the available combustible material in relation to the amount of oxygen required to consume them entirely, by evaluating the chemical combustion reactions in the fire. Methods for this are explained in detail in [15, 20].

4.4.3.2   Smoke Control by Ventilation

A fire releases heat, toxic gases, and smoke. Smoke, microscopic particles suspended in air, is carried by hot air and combustion gases and will raise by buoyancy to the tunnel ceiling in a fuel-controlled fire. At the location of the fire, turbulence mixes air, gases and smoke, and in some distance, from the fire, a layered structure is established (Fig. 4.12). With increasing distance from the fire, the smoke and gases will eventually cool and fill the whole tunnel cross section. Tunnel ventilation with a speed equal or higher than a specific "critical velocity" can keep the part of the tunnel which is upstream from the fire essentially smoke free and allow safe evacuation and access by fire fighters from this direction.

The critical velocity depends on the tunnel geometry (cross sectional area) and on the characteristics of the fire, principally its heat release rate, which can be evaluated from the amount of combustible material and a fire development scenario.

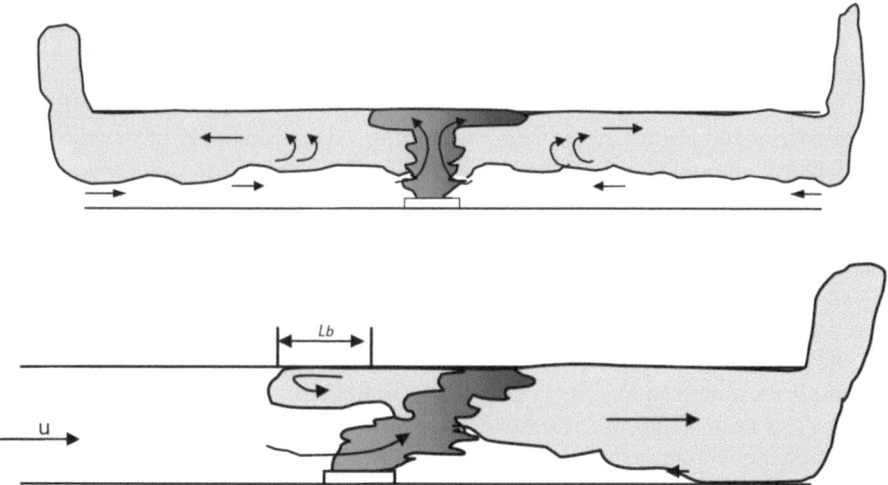

Fig. 4.12 Sketches of smoke stratification during a tunnel fire. Top: low ventilation velocity (0–0.5 m/s). Bottom ventilation speed close to the critical value. From [21]

The design of the tunnel ventilation can be considered as an important part of the fire prevention strategy in an accelerator tunnel. Nowadays the engineering process is supported by numerical fire simulation programs. In simple cases, a so-called compartment model may be sufficient, for more complicated situations a computational fluid dynamics (CFD) code is necessary. The "industry standard" of fire and smoke propagation CFD codes is the Fire Dynamics Simulator from the National Institute for Standards and Technology NIST [23].

4.4.4   Fire Prevention

Going back to the fire triangle, fire prevention focusses on limiting the amount of combustible material and on controlling ignition sources. Only exceptionally, the oxygen content of air can be reduced to prevent fires, for example in archive- or computer buildings. Fire prevention consists of evaluating the fire risk, taking mitigating measures, detecting a fire, and securing the evacuation of personnel.

4.4.4.1   Fire Risk Assessment

The prevention of fires starts with a fire risk assessment. During this exercise the assessment team, which ideally consists of safety experts, personnel familiar with the facility and line managers, the presence of flammable materials and of ignition sources is determined. This is a specialised from of the hazard register for general risk assessments (Sect. 5.1.2).

In a second step, the probability of fire scenarios from the conjunction of ignition sources and fuel are evaluated. While the probability of electrical failures leading to fires can be taken from industry statistics, the estimation of a probability for human failure is more approximate.

In the last step, the consequences of the most probable fires must be evaluated. Heat release curves for the identified flammable materials inform about energy released in during the fire, this must be put into relation with the fire resistance classes of the building materials. An elaborate consequence analysis can make use of fire and smoke propagation CFD programs [23].

Fire risk analysis is the topic of a series of international standards [22], its third part consists a worked example for an industrial facility. This is the closest one comes in the literature to an accelerator facility.

4.4.4.2   Fire Risk Mitigation

The fire risk assessment informs about the most probable fire scenarios, in relation with the presence of ignition sources and flammable materials. The logical steps of mitigation are

• Elimination of ignition sources
• Removal of flammable material

It has been outlined (Sect. 4.4.2) that the principal ignition sources in accelerators are of electrical and human nature.

Electrical ignition sources are overheating, sparking, and arcing of electrical components. These can be eliminated already at the design stage of an equipment by a sound design of the electrical circuits, and by enclosing the equipment in spark-proof cabinets. The latter step is mandatory in locations where an explosive atmosphere may potentially form, for example where large quantities of liquid fuels or solvents are stored or used. The technical term is ATEX area (from the French ATmosphère EXplosive). In the European Union, these situations are regulated in the ATEX Workplace directive [17] and the ATEX Equipment directive [18].

During operation, defective components are identified either by a malfunctioning device, or can be detected with infrared cameras which show excessive heating.

Human error as an ignition source is combatted with a system of administrative controls, i.e. regulations, procedures and permits. After a local fire risk assessment, concentrating on the absence of flammable material and the availability of extinguishing means, a *fire permit* is given to a worker to perform hot work (welding, brazing, grinding). Workers in these trades must be made aware of the fire risk they may provoke and trained in methods to extinguish a small fire before it spreads out of control. The discipline of the fire permit process is important, only if it is regularly reminded and enforced one can effectively reduce the number of cases of fire caused by human error.

Control of the characteristics and the amount of flammable materials is the second path to minimisation of fire risk. Easy flammable materials shall be replaced with materials which are difficult to ignite. This measure can be applied for example to electrical cables, where ignition resistant insulations exist today. Old cables should be replaced with such newer types when they arrive at the end of their useful lifetime. The same measure applies to a range of other organic materials. Another example are insulating fluids in high-voltage switchgear. They are mineral oils with a low flashpoint, defined as the temperature where the vapour layer over the flammable liquid spontaneously ignites, and should be replaced with synthetic ester-based fluids with higher flashpoint.

The amount of flammable materials in premises, especially where ignition sources are present, shall be regularly controlled. Only materials necessary for the operation of the facility must remain, everything else, for example packaging materials, goods stored in operational area, and waste, must be cleared out to reduce the probability of a fire.

4.4.4.3 Fire Detection and Evacuation

If a fire emerges it is important to detect it as early as possible, to secure the evacuation of personnel, and to start firefighting before the fire is fully developed. Fire detection systems work by two principles, either smoke detection by measuring the opacity by light absorption, or heat detection with infrared detectors. Smoke detectors with ^{241}Am radioactive α-emitters are no longer considered good practice, because of the long half-life of the radionuclide ($t_{1/2} = 432$ years), which made them hazardous waste once the smoke detectors had exceeded their useful lifetime.

In large facilities, multiple fire detectors are combined in a network with a computer-based monitoring application and with local alarm panels, giving an indication of the fire location based on the detector(s) which have been triggered. When a fire alarm is confirmed, usually by the triggering of two independent detectors, the fire brigade is alerted, and an evacuation of the affected building or area started.

The provision of evacuation pathways is prescribed in local regulations, defining the maximum length, minimal width and standardised signposting of paths leading to emergency exits. These regulations are integrated in the contemporary design of surface buildings (offices, laboratories, workshops), but they are often impossible to implement in large accelerator areas, especially when built underground.

Since construction of any building depends on the obtention of a construction permit by the local authorities, the application process is the right moment to develop exceptional measures for the evacuation of personnel from facilities which cannot be built according to the standards.

The exceptional measures will be based on the fire risk assessment of the facility, which must be completed by an evacuation study. The studies must demonstrate the possibility for personnel to evacuate safely from any accessible part of the facility in case of fire. Technical measures to ensure evacuation are:

Fig. 4.13 Examples of emergency path signage (left) and assembly point (right), after [1]. (Image source: https://publicdomainvectors.org)

- Construction of the facility with structural elements resisting the fire heat for a minimal amount of time. Construction materials are classified in fire resistance classes between 30 min and 2 h for normalised design fires.
- Installation of fire and smoke resistant doors along the evacuation pathway, to bring workers already in relative safety before reaching free air. Like the construction materials, these doors are rated for their resistance capabilities.
- Installation of a fire detection system for early warning of personnel and alerting of emergency forces.
- Clearly visible signage of the emergency pathway with standardised signs. In the European Union, signage of hazards and of emergency paths is regulated in a council directive [1] (Fig. 4.13). Entirely based on pictograms, this signage is understood by personnel of any native language.

Simulation codes for fire and smoke propagation codes can help to design an efficient fire detection system and to demonstrate that the evacuation of personnel is possible without exposing them to excessive heat or smoke.

4.5 Occupational Noise

Exposure to high levels of noise has been recognized as an adverse factor for occupational health and personnel must be protected from it (Fig. 4.14). Short-term effects are acute acoustic trauma, affecting the eardrum or the bones in the middle-ear, temporary threshold shift (a reduced sensitivity to certain frequencies of the audible spectrum) and temporary tinnitus (a ringing in the ears which may continue for extended periods after noise exposure). After long-term exposure, threshold shift and tinnitus may become chronic affections. A long-term exposure to noise may also cause stress with the accompanying symptoms of high blood pressure and bad sleep. Noise is also a concern for the public living in the vicinity of research establishments (Sect. 4.6.1).

Fig. 4.14 Obligation sign
to use ear protection, after
[1], (Image source: https://
publicdomainvectors.org)

4.5.1   Noise Measurement

Noise, or sound in general is transmitted by an acoustic pressure wave. The ear is
sensitive to *sound intensity I*, related to the root mean square pressure p_{rms} of the
acoustic wave by

$$I = \frac{p_{rms}^2}{\varrho c}$$

ρ is the density of air and c the speed of sound. Young healthy adults can distinguish minute pressure changes of $p_0 = 20$ µPa, just above the level of thermal noise
in air. The pain threshold of hearing is situated at about 60 Pa. Like other human
senses, hearing has a logarithmic sensitivity scale.

To characterise noise at the workplace or in the environment, *sound pressure
level* (L_p) is measured. L_p is a logarithmic measure of the sound intensity and is the
quantity indicated by sound level meters:

$$L_p = 10\log_{10}\left(\frac{p_{rms}^2}{p_0^2}\right) = 10\log_{10}\left(p_{rms}^2\right) + 94[\text{dB}]$$

Sound level meters are weighting the frequency spectrum with a sensitivity function mimicking the human ear, the standard weighting function is designated "A" in
the reference document IEC 60651 and thus sound level is expressed in units of
dB(A). Sound pressure levels of a few sources are indicated in Table 4.2.

From the definition of L_p follows that an increase of 10 dB corresponds to a ten-
fold sound pressure level, and 3 dB are approximately doubling sound intensity.
Care must be taken when noise from different sources is added, the correct way is
to add the respective values of p_{rms}^2. The addition of two noise sources with equal
sound pressure level at the location of measurement increases the total L_p by
3 dB(A).

A simple method to obtain a rough estimate of instantaneous sound pressure
level is described in Table 4.3. Its purpose is to trigger precise measurements with a
calibrated sound pressure meter if one of the thresholds is likely to be crossed.

Table 4.2 Selected sound pressure levels. From [28], [31]

Activity or environment	Sound pressure level L_p (dB(A))
Pain threshold	140
Pneumatic drill	125
Milling machine at 1.2 m	85
Vacuum cleaner	70
Conversation at 1 m, business office	60
Quiet room	40
Hearing threshold	0

Table 4.3 Simple test to judge if a noise risk assessment is needed [26]

Test	Probable noise level	Conduct risk assessment if noise level is like this for more than:
The noise is intrusive but normal conversation is possible	80 dB	6 h
You must shout to talk to someone 2 m away	85 dB	2 h
You must shout to talk to someone 1 m away	90 dB	45 min

Table 4.4 Required Mitigation for Noise in the European Union [24]

8-h sound exposure level $L_{EX,8h}$ (dB(A))	Mitigation measure
$L_{EX,8h} < 77$ dB(A)	No health risk from noise
77 dB(A) $< L_{EX,8h} < 80$ dB(A)	Optimise sound exposure
80 dB(A) $< L_{EX,8h} < 85$ dB(A)	Protection against noise recommended
85 dB(A) $< L_{EX,8h} < 87$ dB(A)	Protection against noise obligatory
$L_{EX,8h} = 87$ dB(A)	Legal limit of sound exposure for workers

At the workplace, not only the instantaneous noise level, but also the time-integral of sound pressure level is of interest. The total noise exposure is normalized to an 8-h working day $L_{EX,8h}$. [30]. A worker exposed to 85 dB(A) during an activity lasting 4 h and then working in a silent environment for the rest of the day has an 8-h exposure level of 82 dB(A). The evaluation of $L_{EX,8h}$ for a specific worker requires a detailed analysis of his activities, each of which must be accompanied by a measurement of sound pressure level L_p. Finally, the contributions of the different activities must be added and normalized to 8 h. The necessary remedial action depends on the calculated 8 h-sound pressure level (Table 4.4).

Further information about occupational noise, its sources, effects, and prevention can be found in [30]. The calculations described in ISO 9612:2009 are facilitated by a spreadsheet which can be downloaded from UK Health and Safety Executive [27].

4.5.2 Protection Measures Against Noise

From an 8 h-exposure level of $L_{EX,8h}$ = 80 dB(A) on, protection measures are recommended to prevent adverse effects of noise exposition for the worker. They become obligatory at $L_{EX,8h}$ = 85 dB(A). One distinguishes technical measures and personal protection. Technical measures are preferable, they protect all workers indiscriminately:

- Replace the equipment at the origin of noise with a newer, less noisy type.
- Isolate the noise source by encasing it with noise-absorbing materials.
- Isolate the vibrations of the source transmitted to other structures by placing it on isolating floor mats or by using absorbers.

Only where technical measures are impossible or too costly, or where access to the noisy environment is infrequent, workers must be protected with personal protective equipment and clear instructions. The use of this equipment depends on the level of sound pressure that must be reduced to meet the legal requirements. The attenuation of the protector should reduce the sound pressure level perceived by the employee to about 72 dB(A). Overprotection is counterproductive because the worker can no longer communicate or perceive warning signals. Different options are shown in Fig. 4.15.

In environments where sound pressure level exceeds 110 dB(A), more than one protection can be worn to reduce the sound pressure level at the ear to below 85 dB(A). In the EU, all hearing protection should carry the CE mark as an indication that it meets the essential requirements of the Personal Protective Equipment Regulations [25]. The reference [26] contains a detailed guide for the selection of personal hearing protection.

Fig. 4.15 Personal protection equipment against noise. From left: earplugs, individually moulded earplugs, earmuffs. From [26] with permission

4.6 Environmental Impact

Today, technical projects cannot be realised without a strong consideration of environmental protection. Both the public opinion and the legislator demand accelerator facilities where damaging effects on the environment are minimised or compensated for (Fig. 4.16).

Climate change and the fact that the globe has only limited resources must be considered in the design and development of new accelerator facilities. Initiatives have started to reduce the energetic footprint of particle accelerators.

4.6.1 Releases to the Environment

Accelerator centres, as other technical plant, release various substances to the environment, either planned and within authorised quantities, or accidentally, with potentially damaging consequences. An accelerator facility is not a closed system, but it is open to its environment. Effluents can be either liquid and solid, released with water, or gaseous and aerosols released with air. Noise can be a nuisance for the public living in the vicinity of the accelerator's plants.

4.6.1.1 Liquid Effluents

The operation of CERN's LHC for example produces 160 MW heat in the cryogenic plants, normal conducting magnets, and the air-conditioning system. This heat must be removed continuously. A primary cooling water circuit transports the heat energy by heat exchangers from the secondary, equipment specific circuits. The primary cooling water is then circulated to cooling towers where the heat is shed to the environment.

The secondary cooling circuits are in contact with the accelerator equipment. In radiation areas, cooling water can become slightly activated. The activation products of distilled water are short lived with exception of tritium (^{3}H), with a half-life of 12 years but a very low radiological effect on living organisms. Cooling water occasionally transports activated metal corrosion products, with longer half-lives

Fig. 4.16 Warning sign against environmental hazard. (Image source: https://publicdomainvectors.org)

and higher activity concentrations. Before cooling water circuits are drained for maintenance a sample from the circuit is taken and measured for residual radioactivity. It can only be release to the public network when the activity concentration lies below the legally fixed release limit.

Run-off from building roofs, streets and places can be collected in a separate drainage network connecting to natural surface waters (streams and rivers). Accidental pollution of the drainage network is possible by the accidental rupture of tanks or pipes for polluting liquids, often chemical products. A second source for surface water pollution is the inattention of personnel, cleaning polluted tools or vessels and shedding the cleaning water simply outside of the building.

4.6.1.2 Air-Borne Effluents

Environmentally polluting products can be released by air from a particle accelerator centre.

Under certain weather conditions, the steam plumes from cooling towers can be clearly distinguished even from a distant vantage point, and while this may not please the spectator, it does not represent a danger in itself. However, the cooling towers must be regularly treated against legionella. Bacteria from this family thrive in stagnating, lukewarm and represent a hazard for the causation of serious respiratory illness.

The air from accelerator tunnels is activated by secondary particle cascades (Sect. 3.2) and the products from the spallation process are transported to the environment. Most spallation products of pure air are short-lived, with the exception of ^{7}Be with a half-life of 57 days and ^{3}H with 12.3 years. If activated molecules are attached to aerosols, then they can be retained by filters, but gaseous activity cannot be retained. The environmental impact of radioactive releases is estimated with environmental screening models [35], making conservative assumptions about the migration of radionuclides in plants and livestock and the eating habits of the population. In such models the impact CERN's accelerators on persons living in its vicinity has been calculated to be a factor of 10 below the legal limit of $300\mu Sv$/year.

Some technical gases have a high potential to either damage the atmosphere's ozone layer (e.g. fluorinated gases) or they have a high climate potential (e.g. methane). Ozone–layer damaging gases are controlled under the Montreal protocol [37] and they are used under exceptional permission, with the aim to remove them entirely. The use of climate-active gases should be reduced with the efforts to make the industrialised economies CO_2-neutral in a time-span of 20 to 30 years.

4.6.1.3 Environmental Noise

Environmental noise is a by-product of industrial processes which may cause distress and even illness to persons exposed to it [38]. The World Health Organisation recommends an environmental noise level of less than 35 dB in the night to guarantee a reposing sleep. Legally binding limits for environmental noise are decided on national level.

At accelerator facilities, sources of environmental noise are cooling and ventilation plants (fans, blowers), cryogenic plants (compressors), transformers and other power electrical equipment (transformer noise). When these plants are properly designed and protected, environmental noise should not cause problems. As for occupational noise, sources must be isolated from the environment so that vibration and sound cannot propagate. Transformer parks, which cannot be placed in a building, must be surrounded by sound-proof walls.

Many accelerator facilities in Europe were originally built as green-field projects, reasonably far from towns and villages. The growth of urban areas had the consequence that densely inhabited quarters were developed in the vicinity of the accelerator sites. Their inhabitants are exposed to environmental noise from the facility. In these cases, compromises between the local governments, the inhabitants and the accelerator plant's management must be negotiated to achieve a solution which satisfies all parties.

4.6.1.4 Environmental Monitoring

Environmental monitoring serves to demonstrate that the unavoidable release of pollutants from a technical plant is well below the legal limits.

Environmental monitoring takes place at two locations: at the source, where the effluents leave the plant (emission measurement), and in the environment, where one tries to measure the concentration of the substance which has been deposited in the environment (immission measurement).

At the source,

- the activity of radioactive isotopes in released air is measured with special ionisation chambers for gases, and with gamma- or beta-spectrometric methods for the deposits of aerosols on filters in the ventilation release;
- radioactive effluents in water are registered with scintillation detectors immersed in the outflowing water;
- the pollution of water with hydrocarbons can be assessed on-line by catalytic oxidisation or by optical measurements.

In the environment,

- the level of stray radiation escaping the accelerator shielding is measured by ionisation chambers and passive thermo-luminescence dosimeters;
- a potentially radioactive pollution of the biosphere is excluded by spectrometric measurements of media like soil and water and produce like fodder and grain;
- aerosol samplers are placed in some distance to the monitored plant, they aspire large quantities of air which passes a filter, which is measured for radioactive substances after a fixed collection time;

• environmental noise is assessed with sound detectors which are ruggedized for
  environmental, all-weather use.

4.6.2 Reducing the Energetic Footprint

In the last decade, society became increasingly aware of the negative impact of fos-
sil energy consumption on the environment and the climate. High-energy accelerators and colliders built for applied and fundamental research consume large amounts
of electrical energy and it becomes imperative to minimise their energetic and cli-
matic impact.

In Europe, major particle accelerator centres have pooled their resources and
collaborate in the framework of EU research programmes on the development of
strategies to reduce the energetic footprint of accelerator facilities. A series of
workshops summarises the results of this research and development efforts.
[33, 34], in the proceedings of these workshops more details are found to the following examples.

4.6.2.1 Energy Supply from Sustainable Sources

The European Spallation Source ESS in Lund (SE) plans to use only energy from
renewable resources (water, wind, solar) bought on the Swedish electricity market,
on which more than half of the offer is sustainable (mostly water-powered
electricity).

4.6.2.2 Energy Recovery

Accelerator facilities use large amounts of energy for cooling (cryogenic plants,
electromagnets, data centres, air conditioning). Conventionally, the excess heat
from these processes is shed to the environment by heat exchangers and cooling
towers. The excess heat from cooling applications is available at a low temperature level. Heat pumps can convert it to high-temperature heat energy with little
additional energy input. The overall efficiency of such systems is positive. ESS
collaborates with an international energy supplier to build a heat network, supplying parts of the city of Lund with excess heat from the accelerator and target.
CERN will supply excess heat from one of LHC's cryogenic plants to an Eco-
quarter in the nearby town of Ferney-Voltaire. Part of the energy generated
through summer will be injected in a geological storage so that it is available in
the heating period.

4.6.2.3 Accelerator Technology

Elements of the particle accelerator can be optimised to a lower energy consumption.

Under certain conditions, electromagnets can be replaced by permanent magnets. Owing to their fixed magnetic flux density, they are useful in portions of accelerators or beamlines where the energy does not change. In a new linear proton accelerator at CERN (LINAC 4), permanent magnet blocks are used to build a quadrupole magnet in a reduced available space [36]. Adjustable tuning blocks from magnetic steel permit an adjustment of its field gradient within 20% of the nominal value. While space-saving was the technology driver, this application of permanent magnets demonstrates their potential in accelerator technology.

FASER, a forward spectrometer in the LHC tunnel in the for the search for exotic particles [32] employs dipole magnets assembled from permanent magnets to separate charged and neutral particles.

RF cavities can make large gains in energy efficiency when their walls are made from superconducting material for their walls. In a comparison of a hypothetical linear accelerator for a terminal energy of 10 GeV with conventional Cu cavities or s.c. cavities from Nb, R. Porter [in 34] finds that the superconducting solution would use more than 100 times less power than the conventional, and it would be 20 times shorter. Most of the power in the s.c. accelerator goes in the cooling of the cavities to 2 K. Switching to the novel s.c. material of Nb_3Sn, which can work at 4.2 K, would allow a power gain of a factor 400.

As a conclusion to this section of energy efficiency one may remark that particle accelerator technology is a field at the forefront of research and development. It is expected that energy-efficient technologies developed for accelerators will rapidly make their way to applications in other fields of society, as has happened previously with accelerator-related technologies for industry and medicine.

References

1. Council Directive 92/58/EEC on the minimum requirements for the provision of safety and/or health signs at work, (EEC Brussels, 1992), http://data.europa.eu/eli/dir/1992/58/2019-07-26

Electrical Safety

2. European Commission, Summary of references of harmonised standards published in the Official Journal – Directive 2014/35/EU, (Brussels, 2019), https://ec.europa.eu/docsroom/documents/38783

3. Directive 2014/35/EU of 26 February 2014 on the harmonisation of the laws of the Member States relating to the making available on the market of electrical equipment designed for use within certain voltage limits, http://data.europa.eu/eli/dir/2014/35/oj
4. Institut national de recherche et de sécurité, Habilitation électrique, (INRS, 2017), http://www.inrs.fr/risques/electriques/habilitation-electrique.html

Mechanical Safety

5. Directive 2006/42/EC of the European Parliament and Council of 17 May 2006 on machinery, and amending Directive 95/16/EC (recast), http://data.europa.eu/eli/dir/2006/42/2019-07-26
6. EC, Guide to the application of the Machinery Directive 2006/42/EC, Edition 2.2 October 2019, http://ec.europa.eu/docsroom/documents/24722
7. European Commission, Harmonised Standards, https://ec.europa.eu/growth/single-market/european-standards/harmonised-standards_en
8. Health and Safety Executive, Health and safety in engineering workshops, HSE HSG-129, (1999, reprinted 2010), http://www.hse.gov.uk/pubns/books/hsg129.htm
9. Health and Safety Executive, Workplace transport safety - A brief guide, HSE INDG199(rev2) (2013), http://www.hse.gov.uk/pubns/indg199.htm
10. Health and Safety Executive, MSD – Manual handling, http://www.hse.gov.uk/msd/manual-handling.htm
11. Institut national de recherche et de sécurité, Dossier Risque Mécanique, (2015), http://www.inrs.fr/risques/mecaniques/ce-qu-il-faut-retenir.html
12. T. Jepsen, *Risk Assessments and Safe Machinery, Ensuring Compliance with the EU Directives* (Springer, Cham, 2016)

Pressure Vessels

13. International Organization for Standardization (ISO), Cryogenic vessels — Pressure-relief accessories for cryogenic service, ISO 21013 Part 1–4 (Geneva 2008 ff)
14. International Organization for Standardization (ISO), Safety devices for protection against excessive pressure, ISO 4126 Part 1–10 (Geneva 2010 ff)

Fire Safety

15. A. Beard, R. Carvel, *The Handbook of Tunnel Fire Safety* (Thomas Telford Publishing, London, 2004)
16. Directive 92/58/EEC on the minimum requirements for the provision of safety and/or health signs at work (EEC 1992), http://data.europa.eu/eli/dir/1992/58/2019-07-26
17. Directive 1999/92/EC on minimum requirements for improving the safety and health protection of workers potentially at risk from explosive atmosphere (EC 1999) http://data.europa.eu/eli/dir/1999/92/2007-06-27

18. Directive 2014/34/EU on the harmonisation of the laws of the Member States relating to equipment and protective systems intended for use in potentially explosive atmospheres (EU 2014) http://data.europa.eu/eli/dir/2014/34/oj
19. Geasmtverband der Deutschen Versicherungswirtschaft e.V. (GdV) (German Insurance Association). Lithium Batteries, Publication VdS 3103en: 2019-06(03), (Köln 2019), https://shop.vds.de/de/produkt/vds-3103en/
20. M. J. Hurley (ed.), *SFPE Handbook of Fire Protection Engineering* (Springer, New York Heidelberg Dordrecht London), p. 2016
21. H. Ingason, *Tunnel Fire Dynamics* (Springer, New York Heidelberg Dordrecht London, 2015)
22. International Organization for Standardization (ISO), Fire safety engineering — Fire risk assessment, ISO 16732 Part 1–3 (Geneva 2012 ff)
23. National Institute of Standards and Technology, Fire Dynamics Simulator (FDS) and Smokeview (SMV), https://pages.nist.gov/fds-smv/

Noise

24. Directive 2003/10/EC of 6 February 2003 on the minimum health and safety requirements regarding the exposure of workers to the risks arising from physical agents (noise) http://data.europa.eu/eli/dir/2003/10/2019-07-26
25. EU, Regulation (EU) 2016/425 of 9 March 2016 on Personal Protective Equipment (2016), http://data.europa.eu/eli/reg/2016/425/oj
26. Health and Safety Executive (UK), Controlling noise at work (2005) https://www.hse.gov.uk/pubns/books/l108.htm
27. http://www.hse.gov.uk/noise/calculator.htm
28. P. Hughes, E. Ferret, *Introduction to International Health and Safety at Work* (Butterworth-Heinemann, Oxford, 2010)
29. Institut national de recherche et de sécurité, ED 6035 Évaluer et mesurer l'exposition professionelle au bruit (2009) http://www.inrs.fr/media.html?refINRS=ED%206035
30. International Organisation for Standardization, ISO 9612:2009 Acoustics - Determination of occupational noise exposure -- Engineering method (ISO, Geneva, 2009)
31. World Health Organisation, Occupational exposure to noise: evaluation, prevention and control http://www.who.int/occupational_health/publications/occupnoise/en/

Environmental Impact

32. EC, Regulation (EC) No 1272/2008 on classification, labelling and packaging of substances and mixtures, (EC Brussels: 2008) http://data.europa.eu/eli/reg/2008/1272/oj.
33. ESSRI, 4th Workshop "Energy for Sustainable Science at Research Infrastructures", Bucharest-Margurele (HU), 23.-24. 11. 2017. https://indico.eli-np.ro/event/1/
34. ESSRI, 5th Workshop "Energy for Sustainable Science at Research Infrastructures", PSI Villigen (CH), 28.-29.11. 2019. https://indico.psi.ch/event/6754/
35. International Atomic Energy Agency, Safety Reports Series No. 19, Generic models for use in assessing the impact of discharges of radioactive substances to the environment (Vienna 2001)

36. D. Tommasini et al., Design, Manufacture and Measurements of Permanent Quadrupole Magnets for Linac4; 22nd International Conference on Magnet Technology (MT-22) 12–16 September 2011, Marseille, France; IEEE Trans. Appl. Supercond. 22 (2012) 4000704. https://doi.org/10.1109/TASC.2011.2179391
37. United Nations Environment Programme, Ozone Secretariat, https://ozone.unep.org/
38. World Health Organisation, Burden of Disease from Environmental Noise (Geneva 2011), https://www.who.int/quantifying_ehimpacts/publications/e94888/en/
39. S. Mannan (ed.) Lees' Loss Prevention in the Process Industries (Fourth Edition), Chapter 12, (Butterworth-Heinemann 2012)
40. Directive 2014/68/EU of the European Parliament and Council of 15 May 2014 on the harmonisation of the laws of the Member States relating to the making available on the market of pressure equipment, Official Journal of the European Union L 189/ 64 - 259, 27. 6. 2014. http://data.europa.eu/eli/dir/2014/68/2014-07-17

Chapter 5
Safety Organisation at Particle Accelerators

Abstract Particle accelerator centres and research institutions, as other branches of industry and business, bear hazards and risks for accidents and occupational illness for workers. On a larger scale, professional activities may harm the environment and the public living in the vicinity of the plant. After this book has highlighted some of the scientific and technical challenges of occupational safety at accelerators, this closing chapter describes how the safety process works, and how and by which means safety is organised in complex structures. The chapter closes with beam safety and functional safety, two areas where technical hazards are controlled by organisational processes.

5.1 The Occupational Safety Process

Occupational safety shall guarantee the safety of workers, the public and the environment from harm. Harm can manifest itself in form of occupational accidents and occupational illness for the workers, nuisance, and potentially illness for members of the public, and as environmental damage. The prevention of these harmful effects takes the form of

- Reduction of the probability that a harmful event occurs;
- Reduction of the severity of consequences if the harmful event occurs.

For standard industries and services many best-practice solutions for occupational safety exist, based on decades of experience with harmful effects and their mitigation. The management of these facilities can apply proven solutions from similar plants. This is only partially possible in an accelerator facility with complex technical installations making use of leading-edge technology. Frequently, no comparative industry exists from where ready-made solutions for occupational safety can be adopted.

Instead, to develop suitable methods of risk control, as a first step the hazards and risks from a particle accelerator or one of its components or operating procedures must be addressed with help of a seven-step occupational safety process:

1. Definition of scope
2. Hazard register

© The Author(s) 2021

T. Otto, *Safety for Particle Accelerators*, Particle Acceleration and Detection,
https://doi.org/10.1007/978-3-030-57031-6_5

3. Application of Standard Best Practice
4. Risk assessment
5. Definition and implementation of controls
6. Documentation
7. Review

Phases one to five constitute the occupational safety assessment, the last phase inserts occupational safety in the lifecycle of the facility, equipment, or procedure by demanding periodic reviews of the validity of the original assessment. An important result of the safety process is the description of hazards and risks at the accelerator facility, in steps 2 and 4. This permits to identify similar harmful situations in "standard" industries, and to adopt their mitigation strategies, suitably modified where necessary, to the accelerator facility.

The occupational safety process can be conducted in the planning stage after the decision for a new facility, component or operational procedure has been taken. It should also be applied when legacy equipment is taken back into operation, which occurs sometimes in research establishments. The earlier the process is begun, the better are the chances to control hazards *by design.* A safety process team is formed by engineers, physicists and other experts from the design or operational team, complemented by one or more specialists in occupation health and safety.

5.1.1 Definition of Scope

The starting point of each occupational safety assessment is a clear definition of its scope. This can be a single workplace (a mechanical workshop, a chemical laboratory), an equipment (a magnet, a machine tool) or an activity (installing an accelerator magnet, testing a radiofrequency device). In some cases, the safety process covers a whole project, for example a new accelerator facility in the planning stage. Most appropriately, the safety assessment for a large project follows a project breakdown structure and progresses by work packages and their contributions to the whole endeavour.

5.1.2 Hazard Register

The second step of the occupational safety process is the establishment of a hazard register. A hazard register is the result of a systematic identification of activities, equipment, and substances with their associated hazards for workers, public or environment. A systematic list of hazards serves as a reminder to not overlook any danger and unifies the terminology among different assessors. Such lists are available from different occupational safety organisms [e.g. [11]] or can be downloaded from the internet. As they are targeted to manufacturing industries and

Table 5.1 Hazard domains for a hazard register, with examples. A detailed hazard list is given in the Annex.

Hazard Domain	Examples
External	Earthquake, climate, and weather
Environmental	Release of pollutants, energy consumption
Physical	Temperature, noise, electromagnetic fields
Ionising Radiation	Particle beam, stray radiation, activation
Non-ionising radiation	UV light, microwaves, lasers
Noxious substances	Chemically or biologically harmful substances
Fire	Ignition sources, flammable materials
Mechanical	Cutting, crushing, collision, fall of object,
Electrical	Electrical shock, electrical arc
Working conditions	Temperature, lighting,
Physiological	Working posture, vibration, manual handling
Unexpected events	Loss of control, loss of power
Organisation	Constraining schedule, lack of information
Psycho-social	Incomplete and monotonous activities

services, they must be enlarged for particle accelerators, or for research establishments in general, to include hazards unique to these environments. Table 5.1 shows the headlines of a hazard register for particle accelerators and Annex A gives a detailed hazard list, tailored to particle accelerators. The adaptation of the hazard list to the local workplace, by adding hazards specific to equipment, activities, or substances in use, or suppressing superfluous hazards, precedes establishing the register.

Depending on the phase of a project in its lifecycle, a hazard register will assess different subjects. In the planning stage, one will look for hazards which can be controlled by design, for example by adopting standards. In the operational phase, workplace hazards will become more important, for example related to the organisation of work or to physiological constraints. A specific hazard register may be called for if vulnerable persons are employed, for example apprentices with little experience, or physically handicapped individuals. In the end, no two finalised hazard registers for a specific activity or equipment will look the same, and only the identified hazards will be written; the register will thus be more specific and shorter than the full list in the Annex.

5.1.3 Application of Standard Best Practice

For many occupational hazards, standard best practice exists to effectively protect persons, the public and the environment. The measures under this heading are often borne from common sense and they may be easy to apply, making further risk assessment superfluous. Sometimes, national legislation requests the implementation of protective measures once a certain hazard is incurred. Other sources for standard best practice are product documentation from manufacturers, recommendations in international standards and the websites and publications of occupational safety organisms [2, 5, 6]. Finally, every organisation will build up over time a wealth of *Return of experience* from accidents and incidents, helping to avoid similar occurrences. The material constituting standard best practice can be systematically collected and classified in a Safety management system.

An important source of Standard Best Practice, and often obligatory to apply, are legal regulations for hazardous equipment, activities, and practices. They may be codified in national labour law, in regulations from professional bodies or in mandatory requirements from insurance companies. A hazard is often considered "under control" when such obligatory regulations are implemented and applied.

In the European Union, an important body of obligatory regulations are published in the form of European Directives. Numerous types of consumer and industrial products are subject to European Directives, including chapters on Essential Health- and Safety Requirements (EHSR). Their purpose is to guarantee equal safety standards for workers and for consumers (buying and using products) throughout the 27 member states of the Union. Only under this condition, goods can be freely traded among the members. Suppliers from third countries who introduce their products to the common market must also apply the European safety standards. These regulations probably make the EU member states the area on the globe with the highest developed safety conformity regulations. Annex B gives an overview of the European Directives pertaining to technologies and products pertaining to the construction and operation of particle accelerators.

5.1.4   Risk Assessment

Risk assessment is the next refinement stage in the occupational safety process.

It becomes necessary when technologies are employed carrying hazards not covered by standard best practice. In a particle accelerator facility, using built-to-purpose high-tech equipment, this is frequently the case. Parts of accelerator hardware may be designed and built in-house or they are a legacy from a time before the applicability of present-day international standards or directives. In special cases, the application of non-compulsory standard best practice may turn out to be too complicated, or too costly. In all these cases a proper risk assessment must be conducted, including an estimation of the likelihood of the hazard causing harm, and of the severity of the consequences.

A second use of a risk assessment is the management of resources. The safety budget for a facility is spent most effectively at those places, where risks can be either eliminated or mitigated with the highest cost-to-benefit ratio by taking appropriate control measures. A correctly conducted risk assessment will identify the areas where the safety budget is spent with the highest effect.

Depending on the complexity of the hazard's source, a risk assessment can vary from a simple judgement of risk on a scale with three or more levels to a full probabilistic risk assessment, requiring man-years of work by experts and recorded on thousands of pages.

5.1.4.1 Semi-quantitative Risk Assessment

A semi-quantitative risk assessment gauges the size of risk by using a two-dimensional risk matrix. Risk is quantified by grading both the probability of a hazardous event and the severity of its consequences on scales with three to five levels. A simple example is given in Table 5.2.

Risk matrices seem straightforward to apply when one has found agreement with the colleagues participating in the process on a common interpretation of the descriptive terms for the probability and potential severity.

Drawbacks of the semi-empirical method with risk matrices are that unexperienced assessors may find it difficult to judge the level of probability or severity of a hazardous event where no statistical accident data exist. This may lead to a schematic application of the risk matrix, giving the impression of scientific accuracy, while the resulting measure of risk is biased by bad input estimates, leading to under- or overestimation [1].

Since the scales of probability and severity in semi-quantitative risk assessment are non-numeric, they must be adapted to the domain from where the hazard comes. The scale of probabilities for an accident with a standard tool may range from "once a day" for frequent occurrences to "once a year" for unlikely events, whereas the probability scale for accidental releases of environmentally hazardous products may

Table 5.2 Example of a risk matrix (adapted from [3]). Probability of occurrence and severity of harm are graded on a 3-level scale, the product yields a quantified risk, which is symbolised by the degree of shading.

		Potential severity of harm		
		Slightly harmful 1	Harmful 2	Extremely Harmful 3
Probability of harm occurring	Highly unlikely 1	Trivial 1	Tolerable 2	Moderate 3
	Unlikely 2	Tolerable 2	Moderate 4	Substantial 6
	Likely 3	Moderate 3	Substantial 6	Intolerable 9

range over much longer periods, from 1 per year to 10^{-4} per year and even 10^{-6} per year for accidents in hazardous industries with lethal effects in the general population. In practice, one will use a different matrix for each domain of application with adapted scales of probability and severity, which leaves the question how their results compare with each other and how they can be combined.

The difficulties of obtaining reliable estimates of the frequency of adverse events and judging their severity of consequences limits the usefulness of risk matrices and it may turn out to be simpler and more reliable to simply apply standard best practice to control a specific hazard.

5.1.4.2   Quantitative Risk Assessment

In some cases, quantitative risk assessments, quoting failure probabilities and resulting in combined probabilities of accident scenarios are required by law. This is the case in chemical facilities classified as "Seveso III" (after the north-Italian town where 1975 a release of dioxin from a chemical plant contaminated the surrounding towns and villages [4] and in facilities belonging to the nuclear fuel cycle.

A widely introduced method of quantitative risk assessment is Probabilistic Safety Assessment (PSA). This method is based on a functional description of the system which is analysed in form of fault- and event tree diagrams.

A Fault-tree (Fig. 5.1) places a system fault (a state in which the system no longer fulfils its purpose) at the top of the diagram. The causes of the fault extend to the bottom, until the diagram stops at the root causes, which cannot be further reduced. In some cases, two causes have to occur simultaneously to trigger a fault (AND relationship), in other cases they act independently from each other (OR relationship) By assigning a failure probability to every element in the fault tree and by applying the rules of Boolean algebra one can evaluate the overall probability of the fault at the top of the diagram.

Event trees (Fig. 5.2) are similar to fault trees. They are used to analyse the consequences of an event (which may be a fault identified in a preceding analysis). The probability of the root event is given as a probability of occurrence per time unit. Each new branching of the event tree, customarily drawn from left to right, represents an alternative between two scenarios. If the relative probabilities for one or the other outcome are indicated for each branching, then the final consequences have the same dimension as the initiating event.

With complex facility layouts and interdependencies of elements the method is tractable only with help of specialised computers programs, able to perform the bookkeeping of thousands of critical elements, each with an individual failure model and frequency.

A second difficulty lies in the provision of failure probabilities for the elements referenced in an accelerator facility. Such data are published for standard electronic components, and may be available for process equipment which is also used in nuclear or chemical processing plants, but must be estimated for purpose-built

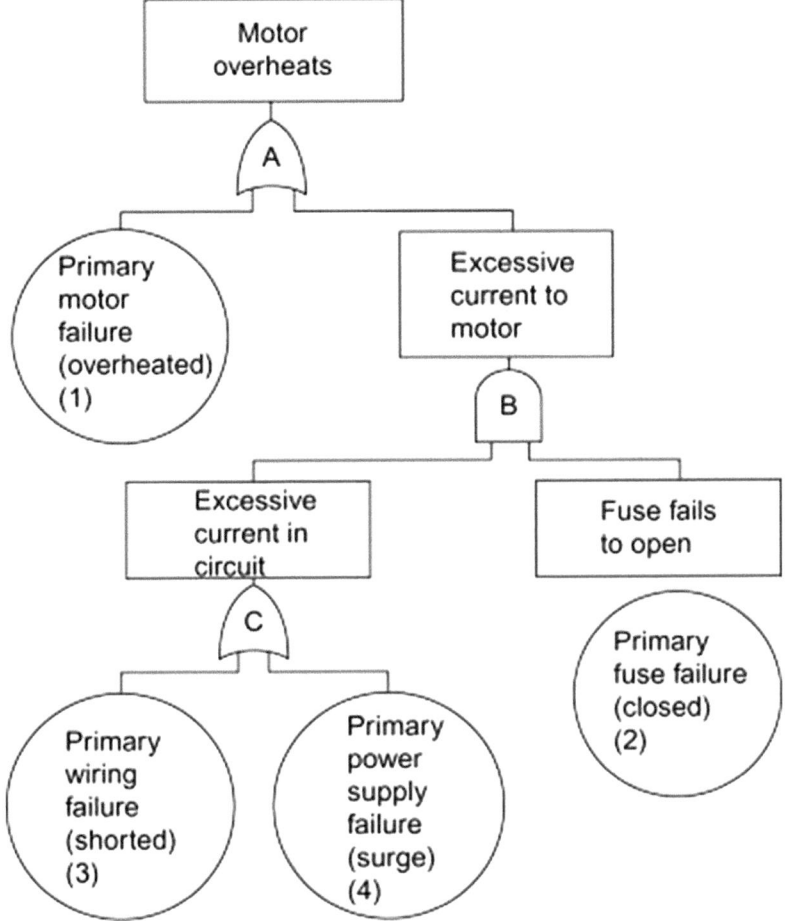

Fig. 5.1 Example for a fault-tree: causes for a motor overheating [7]. (Figure reproduced with permission by Elsevier Science)

hardware components. Special computer programs can evaluate the failure probability pf electronic circuits based on the failure data of the individual components.

   Given the limited potential of particle accelerators to harm persons and the environment in a scale similar to a nuclear power plant or a chemical processing plant, the difficulties obtaining reliable data, and the effort required to set up a full PSA for a complex facility, the method is generally not required for accelerator facilities. Exceptions are accelerators used in nuclear fuel cycle applications, like the transmutation of actinides in the Belgian MYRRHA project [8, 10]. It may also be useful to assess the probability or the consequence of failures of specific, expensive components of an accelerator, for the purpose of protecting it from failures. An introductory treatment of the method with references to further literature is given in [7]

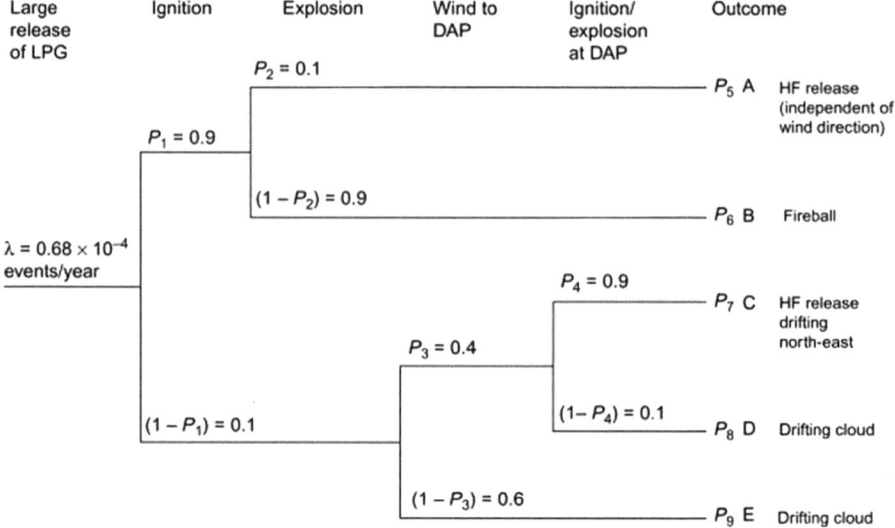

Fig. 5.2 Example for an event tree: consequences of a release of liquid petrol gas [7]. (Figure reproduced with permission by Elsevier Science)

5.1.5 Definition and Implementation of Controls

Once the hazards are identified and their associated risk level estimated, one must decide about controls for these hazards or risks. With control one designates mitigation measures which, once implemented either eliminate a hazard, or render it inoffensive, or at least reduce the risk emanating from it to an acceptable level. After applying controls, a residual risk may remain, which shall be acceptable by the workers and the society.

In the case where compulsory laws or regulations apply, the operator cannot take the decision over the controls, but must conform to the law. Furthermore, if standard best practice is available for the identified hazard, it is often the most efficient way of mitigation.

Often one can identify more than one mitigation measure which would reduce residual risk to an acceptable level, with different effectiveness and cost. Here, the *hierarchy of controls*, going back to the U.S. National Institute for Occupational Safety and Health (NIOSH), gives a decision aid by establishing a hierarchy of mitigation measures, from the most desirable to the just acceptable (Fig. 5.3) [9].

In this hierarchy, personal protective equipment (PPE), and administrative barriers, including safety training, occupy the two lowest levels. This means that they should rarely be the main vector of risk reduction, but they can be useful to control the residual risk after other, more effective controls have been implements, such as elimination and substitution of the risk, or technical protective measures (engineering controls).

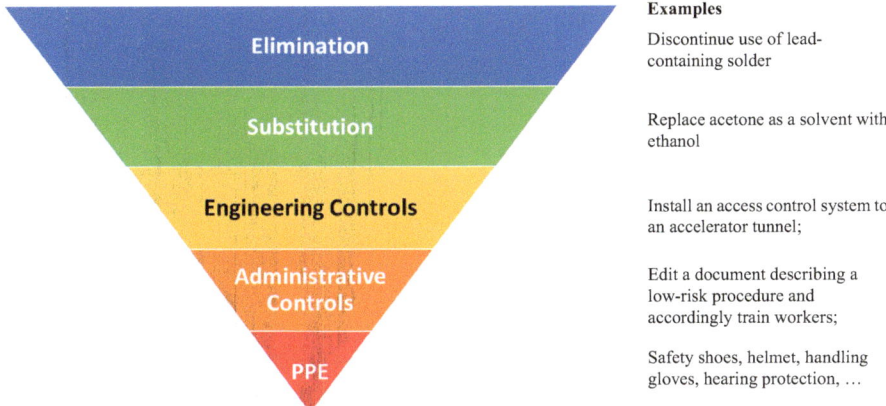

Examples

Discontinue use of lead-containing solder

Replace acetone as a solvent with ethanol

Install an access control system to an accelerator tunnel;

Edit a document describing a low-risk procedure and accordingly train workers;

Safety shoes, helmet, handling gloves, hearing protection, ...

Fig. 5.3 Hierarchy of controls. (Figure according to [9])

5.1.6 *Documentation*

The collected safety-related documentation constitutes part of the memory of the organisation operating the accelerator. It allows to retrace the assessments, judgements and decisions taken. It will ease the construction of similar equipment, for which the same mitigation may be employed. The safety-related documentation may become important to prove the good faith of the organisation, should it be accused of supposed or real damage to workers or to the environment. Documentation forms a core part of the *Safety management system.*

Safety documentation contains also *Return of Experience* (REX). These are the records of inspections, of equipment failures and of accidents, together with a cause analysis. Over time, a body of experience will accumulate and constitute precious information for new facilities or upgrades of the existing ones. The collection of reports concerning failures of accelerator-specific equipment or activities with consequences for safety is particularly important because they are not available from national statistics or accident registers.

Large bodies of documentation should be organised in a document management system to ease future retrieval. If the organisation already uses such a system for the technical documentation of the accelerator facilities, it represents the most suitable location for safety documentation. One can profit from established principles for document classification, and from review, approval, and release procedures. Safety documents relating to a certain piece of equipment should be stored together with technical documentation of this item, simplifying future searches in the archives.

If the organisation has no document management system, then the safety documentation system may proceed by different ways: relative to the equipment, to the organisational subunit, or to the accelerator facility. The important issue is to stick to the choice once made, so that a coherent safety documentation is successively built up.

5.1.7   Review

With hazard and risk assessments completed and mitigation measures implemented, an equipment or a process may start operating and delivering results. In the lifetime of an equipment or a process, there are good reasons to review the safety measures decided previously:

- In case of changes on the equipment/in the process. Here one must assure that the risk assessment remains valid despite the changes, in the contrary case, a revision of the risk assessments and of the resulting mitigation measures is necessary.
- In case of an accident or near miss occurring with the equipment or in the process. This may be a sign that something was overlooked in the original risk assessment or that the mitigation measures were not adequate or not correctly applied.
- The occurrence of accidents in similar facilities or with similar equipment should trigger a re-evaluation of the proper equipment and processes.
- Changed legislation may force to re-assess risks and mitigation measures.
- Finally, a periodic review of all risk assessments which were not reviewed for the reasons above is useful to keep all assessments and mitigation measures updated and adhering to the same standards, even for equipment which never failed within decades of operation.

Reviews inscribe the safety process into the life-cycle of the organisation and are part of its *Safety Management System.*

5.2   Safety Organisation and Management

Institutions operating particle accelerators tend to be large organisations with a hundred or more employees and other collaborators, for example contract workers or research scientists from other institutions. At this organisational scale it is important to clearly define the responsibilities and duties for Occupational Health and Safety (OHS) of every person working on the premises of the institution. A Safety Organisation, based on simple principles and clear rules, gives all concerned parties, managers, employees, contractors and external researchers, guidelines on how to ensure safe working conditions for themselves and others and how to react in case of accidents.

5.2.1   Employer- and Hierarchical Safety Responsibility

A founding principle of OHS is the employer's responsibility for safe and healthy working conditions. This principle is codified in the Occupational Safety and Health Convention, No. 155 [14], to which the member states of the International Labour

Organisation (ILO) mutually agreed in 1981. The ILO has been founded in 1919 as a tripartite organisation with representatives of governments, employers, and workers, with the aim of promoting social justice and internationally recognized human and labour rights. In 1946 it became a specialised agency of the United Nations, it has 187 member states and has its headquarters in Geneva, Switzerland.

Article 16 of ILO Convention 155 stipulates that employers must ensure the safety of workplaces, machinery, equipment, and processes. They must also ensure the absence of risk from chemical, physical, and biological agents when appropriate protection measures are taken, and they must supply the necessary protective equipment to the workers.

In a small enterprise, the employer is readily identifiable, but in a larger organization with a hierarchical management structure, the employer responsibility is usually extended to a model of *hierarchical safety responsibility*. This means that in matters of occupational health and safety, each member of the hierarchy ("manager") acts at his level of responsibility as if he was the employer. At the lowest level of the hierarchy, each worker is responsible for his own safety and of those who may be directly harmed by his activity.

Translated to an accelerator centre the principle means that, for example, a manager in charge of maintenance assures that the workers under his control (employees and contract workers alike) have received the necessary safety instructions and equipment and apply them in the intended way while performing their duties. The workers themselves are required to follow any instructions given to them and to use the protective equipment provided to them by the manager.

In application of the principle of hierarchical responsibility, the higher levels of the hierarchy must provide the necessary means to the lower levels so that these can fulfil their obligations in matter of safety. These means comprise

- Budget and manpower to plan, finance, and implement safety-relevant measures;
- Authority to take decisions;
- Internal safety regulations for situations not covered by standard industrial safety rules and;
- An allowance of time and budget for safety training and information for all managers and workers.

A sensitive subject is the organization of OHS for contract labour and for scientists from other institutions. Both groups are *external labour* with employers from outside of the institution where they exercise a part of their activities. These employers bear a part of the responsibility for their occupational health and safety. The authoritative ILO document on Safety Management Systems, ILO-OSH-2001 [15] recommends, that the organization receiving external labour shall ensure that its safety and health requirements, or at least the equivalent, are applied to contractors and their workers. This may be achieved by communication and coordination between the appropriate levels of the organization and the contractor prior to commencing and during execution of work and should include provisions for communicating hazards and the measures to prevent and control them. The organization may also be required to provide relevant workplace safety and health hazard awareness and training to external labour if they may encounter specific hazards during their activity. Examples at particle accelerators are safety awareness sessions for electrical hazards or ionizing radiation, hazards which are not customarily encountered by most professionals in "conventional" industries.

5.2.1.1 Safety Policy

The International Labour Organization ILO recommends in [15] that the employer, in consultation with the workers and their representatives, should set out in writing a safety policy. This policy should include as a minimum a statement that the organization commits to the following principles and objectives:

- protecting the safety and health of all members of the organization by preventing work-related injuries, ill health, diseases, and incidents;
- complying with relevant national OHS laws and regulations, voluntary programmes, collective agreements on OHS and other requirements to which the organization subscribes;
- ensuring that workers and their representatives are consulted and encouraged to participate actively in all elements of the OHS management system;
- and continually improving the performance of the OHS management system.

The policy shall contain

- a description of the organization of health and safety, with the names and positions of the responsible managers;
- a reminder of the duties and rights of employees, including contractors and temporary personnel;
- the organization's duties towards the wider public, living in the vicinity of the plant, and the environment in general;
- health and safety performance targets which shall be attained by the organization within a set time span.

The safety policy should be specific to the organization and appropriate to its size and its activities. In a particle accelerator centre, the following statements may become part of the safety policy:

- a commitment to equal rights and duties in matters of health and safety for employees, contract labour, students, and guest researchers
- the priority of health and safety objectives over the availability of the beam, be it for commercial, medical or research purposes;
- a commitment to rigorous handling and elimination of the special waste produced by the facility (chemical, radioactive).

The safety policy should not be a dead document filed in an archive, but an expression of the living commitment of the organization to health and safety. It is important that the principles of the policy are lived by the management, as an example to the employees and other personnel on the site.

5.2.1.2 Safety Support Unit

Occupational Safety and Health is a wide field requiring competencies of scientific, technical, organisational, and regulatory nature. In large organizations such as a high-energy particle accelerator centre it has shown useful to bundle the required competencies in a dedicated safety support unit which reports directly to the highest level of management. Alternatively, for a small accelerator, for example a cyclotron for radioisotope production for medical diagnostics, safety support can be given by specialised consultants.

The purpose of a safety support unit is to reinforce the hierarchical line of responsibility by providing the necessary expertise in occupational health and safety to the different levels of management and to the workers. Areas of expertise include for example

- knowledge of international and national regulations and other Standard Best Practice;
- use of hazard- and risk-assessment techniques;
- scientific or technical competencies in specialised areas of OSH, such as fire prevention, ionising radiation protection or chemical safety;
- competencies in the application of national regulations, such as transport of dangerous goods or elimination of waste.

Members of this unit should have a background in the different technical and scientific professions employed in the organization, completed by training and certification in occupational health and safety. Previous working experience in the accelerator or its supporting laboratories and workshops assures a good understanding of the needs and limits of the clients, i.e. all levels of the organization's hierarchy.

5.2.2 Administrative Safety Controls

In the hierarchy of controls (Sect. 5.1.5), administrative controls occupy one of the last places in the ranking. In the overall safety management system of an accelerator facility they have two important roles:

- they set the scene of the implementation of safety with regulations, procedures, and safety training;
- they guarantee safety during exceptional circumstances when the technical safety controls are overridden, with safe systems of work.

5.2.2.1   Internal Safety Regulations

One task of OHS experts in a safety support unit is the drafting of internal safety regulations. These publications comprise, first, the safety policy of the organization, and the principles of its safety organization. Further regulations fall in two categories:

- Specific implementation of international directives, national laws, or other pieces of standard best practice in the organization;
- Regulations to protect workers, the environment and the public from specific hazards which are not covered by any standard best practice.

Internal regulations shall be clearly written and presented in a way that is understood by those who must use them. It is not recommended to publish safety documents in the style of European Directives or national laws. This would not only reproduce the work by the public bodies who edited these regulatory documents, but their style is also not readily understandable by the target audience.

For workers coming from global collaborations, as in international accelerator centres, the use of illustrations and cartoons may be suitable to pass safety advice. The draft of an internal regulation by the OHS experts shall be reviewed by the organization's management and by worker's representatives before being authorized by the general manager or director and published, for example on a dedicated web site of the internal network.

5.2.2.2   Safety Awareness and Training

Safety policy and safety regulations are only effective if they are known and adopted by the employees on all levels of the hierarchy. A channel for the communicating of safety messages are compulsory safety awareness sessions. Their purpose is to familiarise personnel with the safety policy, or with regulations for specific technical domains. In an awareness session, no hands-on practical training is given, and usually no comprehension test is required from the participants Such sessions can be given to all employees of a small unit *in persona,* either by the supervisor, or a safety specialist. A *toolbox talk*, as these sessions are also called, can be part of regular staff meetings. In large accelerator centres, the awareness sessions for general safety topics can be given as a computer-based content which employees can follow at a suitable moment in time.

Other means of creating safety awareness are poster campaigns, the organisation a *safety day*, and placing safety messages in a prominent place in internal communication media

Safety awareness cannot replace formal safety training for the prevention of accidents and protection of workers. Examples for classroom courses with practical exercises are

- Use of personal protective equipment for work at height (climbing harness, security with ropes);
- Use of personal masks for chemical protection, including a fit-test of the mask;

- Use of fire extinguishers;
- Measurement of ionising radiation at the workplace, for example to control surface contamination in a laboratory for radioactive substances
- Use of the electrical lock-out procedure.

Safety training must be given on a level which is accessible to all categories of personnel in a facility. This brings the problem that workers with higher technical education may feel that the level of the training is too low for them. However, this is often a false problem: safety training for ionising radiation protection does not need to explain the physics of the atomic nucleus, as much as training for work at height does not repeat Newton's laws to explain free fall. As long as safety training is focussed on the safety aspects of a topic it can be made interesting and engaging for all participants, no matter their background.

5.2.2.3   Safe System of Work

In the hierarchy of safety controls (Sect. 5.1.5), technical safety measures are favoured: machine covers protect against mechanical accidents and electrical shock, an access safety system (Sect. 5.3.1) prevents unauthorised entrance to potentially hazardous areas. Most technical safety measures are designed for the normal operation of an equipment. They lose their protective function during maintenance and repair. In the above examples, machine covers are removed when a machine fault is investigated, or standard maintenance executed. In this phase, the worker in contact with the dangerous parts must be protected against sudden energisation of the machine, which would expose him to the risk of a mechanical or electrical accident. In Sect. 4.1.2.2 a solution to this problem in the domain of electrical safety has been presented in form of the lock-out process. It draws its name from the procedure to physically lock-out the energy source with a personal padlock. The five steps of a lock-out procedure are:

1. **Identification** of the power source;
2. **Separation** from the power source;
3. **Locking** the separation with a personal padlock;
4. **Verification** that the equipment is without power;
5. **Securing** the equipment, to make its re-start impossible.

The technical means to accomplish each step are different between e.g. electrical appliances and mechanical equipment, but the purpose remains the same: prevent the accidental start of the equipment while workers are in contact with it.

In large plants, to which many particle accelerators belong, the five steps cannot be executed by the maintenance worker alone. Here, a safe system of work with permits to work [12, 13] takes the relay. It is an administrative process which complements the lock-out procedure where a single person does not control each of its five steps. In a distributed electrical powering scheme, where the power sources are far from the equipment powered, and different units in control of source and equipment, the steps 1–3 are usually executed by a member of the powering unit, while 4 and 5 are the tasks of the worker on the equipment.

Before opening the protective covers of an electrical equipment, the maintenance worker will request its separation from the power source. He uses a standardised form for this request, on paper or by a network-based messaging system. After completing steps 1–3, an appointed person in the powering unit certifies to the maintenance worker, using the same standardised form, that the power source has been locked out and secured. Now the maintenance worker can accomplish steps 4 and 5, which ascertain to him that the correct equipment was separated from power, and that it cannot be started accidentally. When the work on the equipment is terminated, the worker must inform the powering unit and only then they may lift the lock-out and re-energise the equipment.

The information of the powering unit to the maintenance worker is sometimes called a work permit, and sometimes the maintenance worker may give a subsidiary work permit to a colleague who is engaged on the same equipment. In either case, all issued permits must be annulled by returning them to their source before the equipment can be powered again.

This process, which is literally preserving integrity and life of the maintenance worker, must be based on a clear, written, safety procedure, known to all participants. Persons must be appointed in written by their supervisors to the key roles in the safe system of work, and they must have received training on the overall procedure and specific to their role. The requests for separation form the source, the certification for lock-out, and the information about the end-of-work must be stated without ambiguity, using pre-printed forms or a network-based messaging system. A well-thought and rigorously maintained safe system of work represents an efficient safety mechanism to protect against accidents during periods where the technical controls are out of order.

5.3 Beam Safety

Several hazards are related to a particle accelerator operating with beam: magnetic fields (associated to high electrical currents), fluids under pressure (generating an oxygen deficiency hazard if released), and ionising and non-ionising radiation. Depending on energy and intensity of the particle beam, some or all of these hazards may have an intolerably high level of risk. Consequently, access is forbidden to the accelerator area and its operation is piloted remotely from a control-room. An accelerator safety system ascertains that persons cannot be harmed by voluntarily or accidentally getting close to the beam. Its functions are:

• Prevent access to the accelerator area during operation.
• Prevent accelerator operation as long as personnel are present in accelerator area.

- If an access is forced during accelerator operation, the accelerator is brought rapidly to a safe state, by turning off the particle beam. The three first elements are fulfilled by the access safety system
- In periods without particle beam, permit access only to authorised personnel. This part is played by the access control system.

An overview of the topic is given in [16, 17].

5.3.1 Accelerator Safety System

The purpose of the Accelerator Safety System (ASS) is to allow operation of the accelerator only when it cannot cause harm, in particular by preventing persons from accessing the vicinity of the accelerator during operation. The ASS consists of the following elements:

- *Barriers* to make the accelerator or its components inaccessible during operation. For low-energy accelerators or for beamlines with low duty factor the barriers may consist of fences. In many cases the radiation shielding has also the role of an access barrier. The highest-energy particle accelerators are built underground and communicate with the surface with a few access shafts. In this case, the barriers are localised at these shafts.
- *Access doors* through the barrier to the accelerator area. In the simplest case, the access door must be locked by the responsible operator before starting the accelerator. In contemporary facilities the doors are equipped with electronic locks operated remotely from a control room. The closed state of the doors is supervised with an interlock switch.
- *Accelerator interlocks* which prevent the particle beam from circulating. Accelerator interlocks can act on different components of the accelerator: they cut the power supply to the particle source, or to a bending magnet which injects the beam into a different area. They can also consist of a beam stopper, a massive metal block with a similar lay-out as a collimator or a beam dump (Sect. 2.5) which is moved into the beam path to stop or to diffuse the particles and make them inoffensive. The status of accelerator interlocks is verified with electronic switches.
- *Interlock keys* are an element of the accelerator interlocks. Every person accessing the accelerator area removes a key from a distribution panel located at the access door. Only when all keys are replaced in the corresponding locks in the distribution panel, the interlock is raised, and the accelerator can be re-started. This prevents accidentally "forgetting" a person and exposing her to the dangers of an accelerator beam.
- *Patrols* of the accelerator area are organised after longer interruptions of service, for example after a shut-down. The patrol is formed by at least two experienced members of the accelerator personnel with a good knowledge of the area. They make certain that no person is present in the accelerator area before releasing it for operation.

Analogous to preventing personnel access, environmental protection consider-ations can enter the scope of the ASS, for example by preventing the release of harmful substances.

The status of all elements of the ASS is transmitted to the control room. Only when all monitored elements are safe for operation, the accelerator can be started. The ASS uses positive logic: if the switch is either open or not functioning, the accelerator is considered unsafe for operation. Today, an ASS can be realised by transmitting the status of the elements by Ethernet or an equivalent data bus and the logical state can be evaluated by a programmable logic controller (PLC). Special Safety PLCs are on the market which embody internal self-checking functions and redundancy. Some local legislations, however, require that the ASS be realised in "hardware", in this case the interlock switches are wired with dedicated signal cables and the logic controller is realised by discrete electronics, without program-ming. The question of the reliability of the ASS is treated in Sect. 5.4.

5.3.2   Access Control System

The purpose of the Access control system (ACS) is in its name: it controls and regu-lates the access of persons to the accelerator area. Permission to enter an accelerator area depend on several prerequisites:

- The person requesting access must have a permission to do so. This permission is linked to the task he or she must execute.
- For accessing areas with particular hazards (electricity, oxygen deficiency, radia-tion), the person must have received adequate safety instructions or training before being allowed to work there.

The ACS is ranking below the ASS: even when the prerequisites above are met, access is only permitted in operational states of the accelerator in which safety risks are tolerable for personnel. This means generally that no beam is accelerated and that its main components (magnets and RF systems) are not powered.

The access to areas with high beam loss (Sect. 3.1) may be linked to the radiation dose rate.

The ACS is materialized by the access doors, which are also part of the ASS (Fig. 5.4). In small facilities, control room personnel supervise and operate the access doors remotely, the access permission can be checked personally because the operators know each other, or they consult lists of authorised personnel. In larger facilities with dozens or hundreds of employees, this method is error-prone and inefficient. Automatic access control systems are now the standard. Personnel carry a badge with barcode or RF identifier to identify themselves at the access door. Their permits can be coded on the card, or the ACS is linked to a database with all permits. When the permits are sufficient, the door opens automatically and gives access to the area.

Fig. 5.4 Access doors of the Proton Synchrotron at CERN. Right, a personal door, letting only one authorized person pass at a time. Left, a material lock, in which the two doors cannot be open at the same time. Centre, the access control system with interlock key panel. (Photo: CERN)

A biometric identification, where the identity of the cardholder is checked with a fingerprint, an image of the iris or a photo-id may be necessary in facilities where theft or sabotage are feared, for example in nuclear facilities where fissile material is handled under regular control by the International Atomic Energy Agency (IAEA) or by EURATOM.

5.4   Functional Safety and Safety Integrity Levels

Functional safety covers situations where the safety of a system or an equipment depends on *safety functions,* realised by electrical, electronic, or programmable-electronic (E/E/PE) means. The safety function acts on demand, after being triggered by a sensor, or manually, and brings the system or equipment into a safe state.

The previous Sect. 5.3 on beam safety showed a few application examples of functional safety at particle accelerators. The accelerator interlocks of the Accelerator Safety System (ASS) employ electrical door contacts as sensors for an attempted entry to the accelerator tunnel. A safety control system (programmable electronics) acts on the sensor information from the door and gives commands via electrical or electronic switches to beam safety elements, hindering the injection of a particle beam as long as the presence of persons in the accelerator area has not been excluded by a patrol. Likewise, the Access Control System (ACS) relies on doors with electronic locks, personal identification with identity badges and the access to a database where access information of duly authorised personnel is stored.

Other examples of functional safety at particle accelerators are quench detection and quench protection systems in superconducting magnets (Sect. 2.2.3), and the early detection of smoke (Sect. 4.4.4) or of oxygen deficiency (Sect. 2.3.3) by sensors, triggering an evacuation alarm.

To fulfil their purpose to guarantee the safety of a system or an equipment, safety functions must work reliably when they are demanded. [18] gives an overview of the topic with a focus on accelerator applications. The fundamental reference are a series of international standards [20, 22], covering the reliability of safety functions. The standard introduces the Safety Integrity Level (SIL), which quantifies the reliability of a safety function on a scale ranging from 1 to 4. A safety system with SIL 4 has the highest probability to satisfactorily deliver its safety function on demand: a safety system with SIL n has an average probability of a dangerous failure PFD upon demand between $10^{-(n+1)} < PFD < 10^{-n}$. Safety functions covering risks having the highest stakes, for example the protection of human lives, must have the highest safety integrity levels SIL.

The safety integrity level for a safety function is evaluated in a special type of risk assessment. One evaluates the probability of an accident under the assumption that safety functions covering this particular risk are not present. Then, the required SIL of these safety functions are determined such that they fill the gap between the failure probability of the system and the targeted maximal probability for an accident. A simplified, hypothetical example illustrates this concept:

The access safety system for a specific room in an accelerator facility consists of a door with an electronic lock, which is shut when the accelerator operates, and, independent from it, a mobile beam dump which can be moved into the beam to block it from the accessible room. The request to the accelerator safety system ASS is to prevent the exposure of a person entering the room accidentally to the beam. The failure of doing so shall be less than $p_{access,beam} < 10^{-5}$. This system is represented in the simple fault tree in Fig. 5.5.

Table 5.3 shows the steps leading to the evaluation of the SIL of the mobile dump. A component with SIL 1, having a failure probability of up to 10% on demand, would not be sufficient. Therefore, the dump control system must meet the requirements of SIL 2.

In [18] a more elaborate example of functional safety for a particle accelerator safety system can be found.

In many practical applications of functional safety, it is too complicated to draw up a complete fault tree. This approach suffers from the often-incomplete knowledge of failure probabilities of components. In such a case, one can determine the required SIL of a safety function with qualitative methods or with hybrid methods combining quantitative and qualitative elements. Part 5 of [22] summarises recommended methods for SIL evaluation.

Fig. 5.5 Fault tree for the simple ASS in the example for SIL determination. The two safety systems, electronic lock, and mobile dump are independent and can be represented as parallel blocks feeding an AND-gate: they must fail both to lead to a complete system failure.

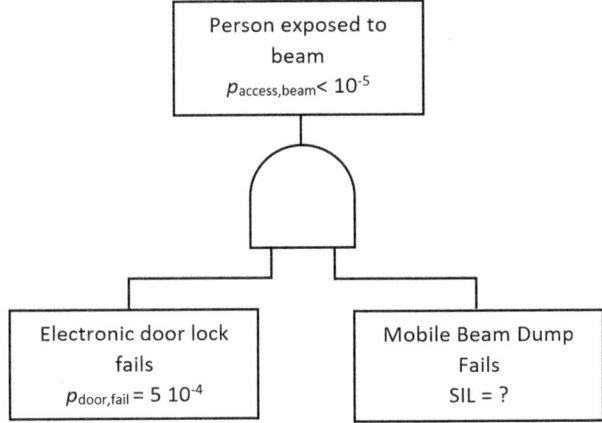

Table 5.3 Failure probabilities of the components of a hypothetical accelerator safety system (ASS) and evaluation of the required SIL of a component

Target failure probability: person enters accelerator room	$p_{access,beam}$	$< 10^{-5}$ per demand
Failure probability of electronic lock	$p_{door,fail}$	$5 \cdot 10^{-4}$ per demand
Required failure probability of moveable beam dump	$p_{dump,fail} = \dfrac{p_{access,beam}}{p_{door,fail}}$	$< 2 \cdot 10^{-2}$ per demand
Average failure probability of a control system with SIL 1	PFD (SIL 1)	$10^{-2} < PFD$ (SIL 1) $< 10^{-1}$
Average failure probability of a control system with SIL 2	PFD (SIL 2)	$10^{-3} < PFD$ (SIL 2) $< 10^{-2}$

Once the required SIL of a safety function is determined, it needs to be implemented. Manufacturers offer electrical and electronic components as well as mechanical components which meet the reliability requirements of a defined SIL. These components have been tested and certified by authorised institutes. For example, in Germany, VDE offers a certification service for functional safety [24].

The concepts of functional safety are also applied to the controls of machines (Sect. 4.2.2) in [21] and to the process industry [23]. The latter standard can be used as a template for the implementation of functional safety in cryogenic systems.

For further information on the topic, the UK Health and Safety Executive maintains an informative website on functional safety [19] with many references to in-depth articles and studies.

References

Safety Process

1. L.A. Cox, What's wrong with risk matrices ? Risk Analysis **28**(2) (2008)
2. Deutsche Gesetzliche Unfallversicherung – Spitzenverband. Prävention https://www.dguv.de/de/praevention/index.jsp
3. Health and Safety Executive (United Kingdom), *Risk Management- Frequently Asked Questions – What are risk matrices?* http://www.hse.gov.uk/risk/faq.htm#q27
4. Health and Safety Executive, ICMESA chemical company, Seveso, Italy. 10th July 1976, http://www.hse.gov.uk/comah/sragtech/caseseveso76.htm
5. Health and Safety Executive, Guidance, http://www.hse.gov.uk/guidance/index.htm
6. Institut national de recherche. et de sécurité, Démarches de prevention, http://www.inrs.fr/demarche/
7. S. Mannan (ed.) Lees' Loss Prevention in the Process Industries (Fourth Edition), Chapter 9 (Butterworth-Heinemann 2012)
8. MYRRHA, Multi-purpose hYbrid Research Reactor for High-tech Applications. https://www.myrrha.be/
9. National Institute for Occupational Safety and Health (NIOSH), Hierarchy of Controls, https://www.cdc.gov/niosh/topics/hierarchy/#
10. B. Neeradael, The MYRRHA Project – Safety methodology and challenges, Plenary contribution at EUROSAFE Forum 2012, https://www.eurosafe-forum.org/eurosafe2012#Plenary
11. SUVA, Connaissez-vous le portefeuille des phénomènes dangereux dans votre entreprise (Luzern 2011), https://www.suva.ch/fr-CH/materiel/documentation/connaissezvous-le-potentiel-des-phenomenes-dangereux-dans-votre-entreprise%2D%2Dle%2D%2D66 105-f-25527-25526

Safety Organisation

12. Health and Safety Executive, Isolation and permits to work, https://www.hse.gov.uk/safemaintenance/permits.htm
13. Health and Safety Executive, Human factors: Permit to work systems, https://www.hse.gov.uk/humanfactors/topics/ptw.htm
14. International Labour Organization, C155 – Occupational Safety and Health Convention (Geneva 1981), https://www.ilo.org/global/standards/subjects-covered-by-international-labour-standards/occupational-safety-and-health
15. International Labour Organization, Guidelines on occupational safety and health management systems, ILO-OSH 2001, 2nd ed. (Geneva 2009) https://www.ilo.org/safework/info/standards-and-instruments/WCMS_107727

Beam Safety

16. R. Schmidt, in Proceedings of the Joint International Accelerator School: Beam Loss and Accelerator Protection, Newport Beach, United States, 5–14 November 2014, edited by R. Schmidt, CERN-2016-002 (CERN, Geneva, 2016), pp.1–20, doi:10.5170/CERN-2016-002.1
17. R. Schmidt, in *Proceedings of the Joint International Accelerator School: Beam Loss and Accelerator Protection*, Newport Beach, United States, 5–14 November 2014, edited by R. Schmidt, CERN-2016-002 (CERN, Geneva, 2016), pp. 319–341, https://doi.org/10.5170/CERN-2016-002.319

Functional Safety and Safety Integrity Levels

18. E. Carrone, in Proceedings of the Joint International Accelerator School: Beam Loss and Accelerator Protection, Newport Beach, United States, 5–14 November 2014, edited by R. Schmidt, CERN-2016-002 (CERN, Geneva, 2016), pp. 271–302 doi:10.5170/CERN-2016-002.271
19. Health and Safety Executive, Functional safety, https://www.hse.gov.uk/eci/functional.htm
20. International Electrotechnical Commission, Functional safety of electrical/electronic/programmable electronic safety-related systems, Part 0: Functional safety and IEC 61508; IEC 61508-0 (Geneva 2005)
21. International Electrotechnical Commission, Safety of machinery – Functional safety of safety-related electrical, electronic, and programmable electronic control systems, IEC 62061 (Geneva 2005)
22. International Electrotechnical Commission, Functional safety of electrical/electronic/programmable electronic safety-related systems, Part 1 – 7; IEC 61508-1 to IEC 61508-7 (Geneva 2010)
23. International Electrotechnical Commission, Functional safety – Safety instrumented systems for the process industry, Part 1-3, IEC 61511 Part 1 to 3 (Geneva 2016)
24. VDE Prüf- und Zertifizierungsinstitut, Funktionale Sicherheit – Prüfung und Zertifizierung im VDE-Institut, https://www.vde.com/tic-de/dienstleistungen/funktionale-sicherheit

Annexes

Annex A: Hazard List for Accelerator Facilities

As described in Sect. 5.1.2, the establishment of a hazard register is the second phase of the occupational safety process for an accelerator facility, one of its components or one of its operational processes. The following systematic list of safety hazards can be edited depending of the scope of the safety process. The list is originally based on a model from the Swiss Occupational Accident Insurance SUVA [1], tailored for accelerator facilities and in use in CERN's Technology Department.

Hazard	Examples
1 External Hazard	
Climate and weather	Storms, heat waves
Earthquake	In sensitive areas, e.g. Italy, Japan, California
Fire, external	
Flooding	If facility close to rivers or at the sea
Ground Pressure	For underground structures (tunnels and caverns)
Landslide	In sensitive areas
Wildlife	
2 Hazard to the Environment	
Activation of ground	Radioactive isotopes ^{22}Na, ^{3}H
Additional Traffic	Commuting of employees, Transport of goods to accelerator sites
Noise (Environment)	Permanent, intermittent, and instantaneous
Release of pollutants: air	Atmospheric emissions (ex: greenhouse gases)
Release of pollutants: solid	Industrial waste
Release of pollutants: water	Impact on the aquatic environment
Energy	Overall consumption of electrical energy
Release of radioactive liquid	By cooling water

© The Author(s) 2021
141
T. Otto, *Safety for Particle Accelerators*, Particle Acceleration and Detection,
https://doi.org/10.1007/978-3-030-57031-6

Hazard	Examples
Release of radioactive solid	
Release of radioactivity by air	By the ventilation system
Waste	Conventional and hazardous waste
3 Physical Hazards	
Field, electrical	EMF, non-ionising radiation
Field, magnetic	EMF, non-ionising radiation
Noise (Workplace)	Workshops, machine rooms
Oxygen deficiency	Accidental release of cryogens
Temperature: cold gas/liquid	In cryogenic systems
Temperature: Heat radiation	
Temperature: Surface, cold	In cryogenic systems
Temperature: Surface, hot	During welding or soldering operations
Ultrasound, Infrasound	Vibrations
Under- or overpressure	
4 Ionising Radiation	
Activated air or gases	Activation of air by stray radiation in accelerator tunnel
Activated dispersed solids	Radioactive dust from machining activated solids
Activated or contaminated liquids	Activated and contaminated cooling water
Activated solids	Solids interacting with beam and stray radiation
Naturally occurring radioactive materials	
Particle beam	Direct exposure to beam in experimental area
Radioactive aerosols	
Radioactive surface contamination	
Radioactive test sources	Used for test and calibration of detectors
X-ray (parasitic)	From RF amplifiers and cavities
X-ray generators	Calibration of dosimeters, structural analysis
5 Non-ionising Radiation	
Laser	Surface treatment, optical measurement, ionisation
Microwaves	
Radiofrequency	From RF amplifiers and wave guides
UV	
6 Noxious Substances	
Dust, particles (non-radioactive)	From machining; nanoparticles
Substance, biologically hazardous	Legionella bacteria
Substance, carcinogenic, mutagenic, sensitising	Benzene, beryllium, cadmium, asbestos, ….
Substance, corrosive	Strong bases and strong acids
Substance, explosive	Inorganic azides
Substance, harmful for environment	Hydrocarbons,
Substance, harmful, irritant	Alkalines
Substance, oxidising	Oxygen, hydrogen peroxide, halogens
Substance, toxic	Carbon monoxide, methanol, arsenic acid,
Cryogenic liquid	Helium, Nitrogen, Argon

Hazard	Examples
7 Fire Hazards	
Combustible materials	Paper, wood, plastics
Explosive atmosphere	Use of large quantities of solvents
Explosives	Flammable gases in cylinders
Ignition source	Open flame, electricity, welding activities
Substance, flammable, pyrophoric	Fuel, solvents, liquefied gas
8 Mechanical Hazards	
Dangerous surface	Angles, corners, cutting edges, roughness
Fall of object from height	During transport with overhead crane or forklift
Fluid under pressure	Gas, vapour, oil, accumulators
Uncontrolled object in motion	Rolling, sliding, projection
Unprotected element in movement	Crushing, shear, cutting, entanglement, abrasion, drawing-in
Yield under stress or pressure, rupture	Material fatigue,
Moving transport and lifting equipment	Cars, slings, lifting chains, cranes,
9 Electrical Hazards	
Electrical contact, direct	Touching live conductor
Electrical contact, indirect	Touching badly earthed metallic surface in accidental contact with live conductor
Electrostatic phenomena	Electrostatic discharge, spark
Short-circuit, overload, electric arc	Arc flash, overheating of ohmic resistance
10 Working Conditions	
Climate in closed environment	HVAC systems, pressurised areas
Climate, weather	For external work only
Heat, cold	
Insufficient lighting	Natural and artificial lighting
Restricted visibility	Fog, smoke
Slippery surface	Spills, effect of snow, ice or rain, wet leaves
11 Physiological Constraints	
Manual lifting and handling	Unloading cars, transport of components
Movement under constraint	RSI (Repetitive Strain Injury)
Repetitive actions	
Vibrations	Hand-arm vibration, whole body
Working posture	Musculoskeletal problems
Confined spaces	Work inside cryogenic tank
12 Unexpected events	
Loss of containment	Rupture of storage vessel,
Loss of control	Failure of electronic- or programmable electronic control system
Loss of cooling	Rupture of cooling water line
Loss of power	Interruption of electricity supply
Protection system failure	Failure of access safety system

Hazard	Examples
13 Organisation	
Constraining schedule	Shift work, irregular schedules
Disorder	Insufficient housekeeping
Frequent interruptions	
Isolated worker	Executing dangerous tasks alone
Lack of clarity of task attribution	
Lack of collaborator participation	
Lack of feedback	
Lack of information and training	No or poorly documented procedures, no training, language barrier
Lack of qualification	Assignment of tasks beyond personal capability
14 Psychological Constraints	
Incomplete and monotonous activities	High degree of work partition
Lack of decision margin	Assembly chain work
Overload	Permanent stress, concentration, responsibility overload, over or underqualification
Relations with colleagues and hierarchy	Discrimination, mobbing, harassment
Emotional charge	

Reference

[1] suvapro, Connaissez-vous le potentiel des phénomènes dangéreux dans votre entreprise ?, SUVA 66105.f, (Luzern 2011), https://www.suva.ch/66105.f

Annex B: European Directives

To strengthen the cohesion between its member states, the European Parliament and Council publish Directives which must be converted into national law within a delay of a few years after publication. The purpose of the "consumer" directives is to provide customers in the whole Union with products meeting agreed safety standards, laid down in the "Essential Health and Safety requirements" of each Directive. The directives formalise a declarative regime, in which manufacturers declare by affixing the "CE" sign on their products that they are conforming to the requirements. For a few, high hazard domains, conformity must be asserted by independent testing bodies. These bodies are "notified", meaning that they operate under a license from a member state and meet requirements of technical competency and quality assurance. One domain where conformity assessment by a notified body is required are pressure vessels, and if one plans to include such vessels in the accelerator design one has to provide for sufficient time and budget for the approval process.

The EU legislation applies also to imported devices. This is relevant if a European accelerator centre acquires specialised equipment from overseas of which the manufacturer (often another laboratory) has not declared the conformity with EU regulations. In this case, the importer must guarantee that such equipment meets the requirements of the directive and its standards. This principle is of course valid in both directions and an accelerator laboratory in the United States has to demonstrate that every piece of equipment acquired from a European or Asiatic manufacturer would meet US standards.

In the *New Legislative Framework* of the EU [4], the formal requirements for conformity declaration are streamlined. An important element is the publication of *Harmonised Standards* [5] by the European standardisation organisations CEN (European Committee for Standardization, [1]), CENELEC (**European Committee for Electrotechnical Standardization [2]) and ETSI** (European Telecommunications Standards Institute [3]). A product designed and constructed in accordance with one of the harmonized standards is deemed to meet the EHSR of the corresponding directive.

European texts with relevance to particle accelerators are:

Transportable Pressure Equipment – Directive 2010/35/EU
Construction products – Regulation (EU) No 305/2011
Simple Pressure Vessels – Directive 2014/29/EU
Electromagnetic Compatibility – Directive 2014/30/EU
Lifts – Directive 2014/33/EU
ATEX – Directive 2014/34/EU
ATEX Equipment Directive 1992/92EC
Low Voltage – Directive 2014/35/EU
Machinery Directive 2006/42/EC
Pressure equipment – Directive 2014/68/EU
Personal protective equipment – Regulation (EU) 2016/425

An overview of the directives in the New Legislative Framework with links to the original texts and further information can be found at [4]

References

[1] European Committee for Standardization, https://www.cen.eu/
[2] European Committee for Electrotechnical Standardization, https://www.cenelec.eu
[3] European Telecommunications Standards Institute, https://www.etsi.org/
[4] European Commission, New Legislative Framework, https://ec.europa.eu/growth/single-market/goods/new-legislative-framework_en
[5] European Commission, Harmonised Standards, https://ec.europa.eu/growth/single-market/european-standards/harmonised-standards_en

Index

A
Accelerator interlocks, 133
Accelerator safety system (ASS), 133, 134
Access control system (ACS), 134, 135
Activation of matter, 62
ATmosphère EXplosive (ATEX) area, 103

B
Bonner sphere spectrometer, 74

C
Carnot factor, 22
Cascade, electromagnetic, 57, 64
Cascade, hadronic, 59, 64
Cold burns, 24
Collider, 10–11
Control, 2–3, 124
Critical energy, 57
Cyclotron, 7–8

D
Displacement, atomic, 61
Dose, absorbed, 65
Dose equivalent, 66
Dosimeter, 67
Dosimetric quantities for non-ionising
    radiation, 36

E
Effective dose, 66
Effect level, 36

Effluents, air-borne, 110
Effluents, liquid, 109, 110
Electrical arc, 20
Electrical hazards of magnets, 19, 20
Electric shock and burns, 84
EMF, Biophysical effects of, 35, 36
Emission measurement, 111
EU Directive, conformity with, 91, 92
EU Directive, Machinery, 90, 91
EU directives, 144–145
Evacuation, 104, 105
Event tree, 122
Exemption level, 79
Experimental targets, 49
External labour, 127

F
Fault-tree, 122
Fire detection, 104, 105
Fire dynamics in tunnels, 101
Fire permit, 103
Fire risk assessment, 102, 103
Fire risk mitigation, 103, 104
Free-Electron Laser (FEL), 10
Functional safety, 135

G
Geiger-Müller counter, 69

H
Harmonised standard, 91
Hazard, 1, 117

© The Author(s) 2021
T. Otto, *Safety for Particle Accelerators*, Particle Acceleration and Detection,
https://doi.org/10.1007/978-3-030-57031-6